Buried Pipe
Design

Buried Pipe Design

A. P. Moser, Ph.D.
Mechanical Engineering
Utah State University
Logan, Utah

McGraw-Hill

New York   St. Louis   San Francisco   Auckland   Bogotá
Caracas   Hamburg   Lisbon   London   Madrid   Mexico
Milan   Montreal   New Delhi   Paris
San Juan   São Paulo   Singapore
Sydney   Tokyo   Toronto

Library of Congress Cataloging-in-Publication Data

Moser, A. P. (Alma P.)
  Buried pipe design / A. P. Moser.
    p. cm.
  Includes bibliographical references.
  ISBN 0-07-043490-5
  1. Underground pipe lines—Design and construction. I. Title.
TJ930.M698 1990
621.8′672—dc20   89-48541

    3 4 5 6 7 8 9 0   DOC/DOC   9 5 4 3 2

ISBN 0-07-043490-5

*The sponsoring editor for this book was Joel Stein, the editing
supervisor was Valerie A. Rothlein, the designer was Naomi Auerbach,
and the production supervisor was Suzanne W. Babeuf. It was set in
Century Schoolbook by the McGraw-Hill Professional & Reference
Division composition unit.*

Printed and bound by R. R. Donnelley & Sons Company.

*For more information about other McGraw-Hill materials,
call 1-800-2-MCGRAW in the United States. In other
countries, call your nearest McGraw-Hill office.*

Contents

Preface

All too often our vitally important underground fluid transportation systems are overlooked or given only cursory review by college texts dealing with water and wastewater treatment or soil mechanics. This text is written to provide the engineering student and the practicing engineer with information basic to the proper, cost-effective design of buried pipe systems. It is a natural outgrowth of many years of study, testing, and research in general areas of pipe performance and pipe installation design. More specifically, this text focuses on the design engineering of buried water systems and sewage collection systems. The varieties of pipe products used in such systems are addressed and examined. Thus, the text may serve as a useful supplement to either undergraduate or graduate courses dealing with such subjects as wastewater collection and treatment, water transmission treatment and distribution, environmental engineering, sanitary engineering, soil mechanics, buried structures, pipeline engineering, and so forth. Because this book contains general design information as well as information on the design of various piping products, it will serve as a valuable resource to the practitioner involved in buried piping system design.

In Chapter 2, methods for calculating soil loads are given. Marston's theory for loads on buried conduits is discussed along with the various factors which contribute to these loads. Underground pipes are placed in tunnels, buried under highways, buried under railways, and buried under airports. Methods are given for the determination of loads that are imposed on pipes in these and other applications.

Design methods that are used to determine an installation design for buried gravity flow pipes are described in Chapter 3. Soil types and their uses in pipe embedment and backfill are discussed. Design methods are placed in two general classes—rigid pipe design and flexible pipe design. Pipe performance limits are given and recommended safety factors are reviewed. The finite element method for design of buried piping systems is relatively new. The use of this powerful tool is increasing with time. A detailed discussion of this method is included.

Chapter 4 deals with the design methods for buried pressure pipe installations. Included in this chapter are specific design techniques for various pressure piping products. Methods for determining internal loads, external loads, and combined loads are given along with design bases.

Chapter 5 deals with various generic pipe products in two general classes: rigid pipe and flexible pipe. For each product, selected standards and material properties are listed. The standards are from organizations such as the American Water Works Association (AWWA) and American Society for Testing and Materials (ASTM). Actual design examples for the various products are given in this chapter.

Acknowledgments

In the preparation of this book, the author has drawn from his previous publications and his lecture notes from the Piping Systems Institute (a short course at Utah State University). Also, source material is used from various standards and handbooks. Acknowledgment is given throughout the book where this material is used.

I express my deepest appreciation to those who helped to make this book possible:

To Uni-Bell PVC Pipe Association and the Vinyl Institute who recognized the need for a better understanding of piping system design and gave encouragement and support.

To Robert Walker, Dave Eckstein, and Dennis Bauer of the Uni-Bell PVC Pipe Association who provided valuable suggestions and timely assistance.

To Dr. R. K. Watkins of Utah State University, R. J. Adams of University of Toronto, and M. A. Collins of Southern Methodist University who reviewed the manuscript of the book and provided constructive criticism.

To the many publishers who graciously gave permission to use their material.

To the Mechanical Engineering secretaries at Utah State University for helping with various aspects of the project but particularly in protecting me so I could spend time writing.

And last, but not least, to my wife Kay for typing and editing the manuscript.

A. P. Moser

Buried Pipe
Design

Introduction and Overview

Underground conduits have served to improve man's standard of living since the dawn of civilization. Remnants of such structures from ancient civilizations, have been found in Europe, Asia, and even in the western hemisphere where some of the ancient inhabitants of South and Central America had water and sewer systems. These early engineering structures are often referred to as examples of the art of engineering. Nevertheless, whether art or science, engineers and scientists still stand amazed at these early water and sewer projects. These seem to bridge the gap between ancient and modern engineering practice. The gap referred to here is that period known as the "dark ages" where little or no subsurface construction was practiced—a time when most of the ancient "art" was lost.

Today, underground conduits serve in diverse applications such as sewer lines, drain lines, water mains, gas lines, telephone and electrical conduits, culverts, oil lines, coal slurry lines, subway tunnels, and heat distribution lines. It is now possible to use engineering science to design these underground conduits with a degree of precision comparable with that obtained in designing buildings and bridges. In the early 1900s, Anson Marston developed a method of calculating the earth load to which a buried conduit is subjected in service. This method, the Marston Load Theory, serves to predict supporting strength of pipe under various installation conditions. M. G. Spangler, working with Marston, developed a theory for flexible pipe design. In addition, much testing and research have produced quantities of empirical data which also can be used in the design process. During the past decade, digital computers combined with finite element techniques and sophisticated soil models have given the engineering profession another design tool which will undoubtedly produce even more precise designs.

Engineers and planners realize the subsurface infrastructure is an absolute necessity to the modern community. It is true we must "build down" before we can "build up." The underground water systems serve as arteries to the cities and the sewer systems serve as veins to carry off the waste. The water system is the lifeblood of the city, providing culinary, irrigation, and fire protection needs. The average man or woman on the street takes these systems for granted being somewhat unaware of their existence unless they fail. In the United States today, people demand water of high quality to be available, instantaneously, at their demand. To insure adequate quality, the distribution systems must be designed and constructed so as not to introduce contaminants.

Sewage is collected at its source and carried via buried conduits to a treatment facility. Treatment standards and controls are continually getting more stringent and treatment costs are high. Because of these higher standards, the infiltration of ground or surface water into sewer systems has become a major issue. In the past, sewer pipe joining systems were not tight and permitted infiltration. Today, however, tight rubber ring joints or cemented joints have become mandatory.

Even though septic tanks and cesspools are still widely used today, they are no longer accepted in urban or suburban regions. Only in the truly rural (farm) areas are they sanctioned by health departments. Today more sewer systems are being installed. This produces a demand for quality piping systems. Thus the need for water systems that deliver quality water and for tight sanitary sewers has produced a demand for high quality piping materials and precisely designed systems that are properly installed.

Soil Mechanics

Various parameters must be considered in the design of a buried piping system. However, no design should overlook pipe material properties or the characteristics of the soil envelope which surround the pipe.

The word soil means different things to different people. To engineers, soil is any earthen material excluding bedrock. The solid particles of which soil is composed are products of both physical and chemical action, sometimes called weathering of rock.

Soil has been used as a construction material throughout history. It is used for roads, embankments, dams, and so forth. In the case of sewers, culverts, tunnels and other underground conduits, soil is important not only as a material upon which the structure rests, but also as a support and load transfer material. The enveloping soil transfers surface and gravity loads to, from, and around the structure. Much has been written about soil mechanics and soil structure interaction. Such variables as soil type, soil density, moisture content, and depth of the installation are commonly considered. If finite element analysis is used, many soil characteristics are required as input to the mathematical soil model. These soil properties are usually determined from triaxial shear tests.

Standards organizations such as The American Association of State Highway and Transportation Officials (AASHTO) and the American Society for Testing and Materials (ASTM) issue standard test methods for classifying soil and for the determination of various soil properties. Of the various methods of soil classification, the Unified Soil Classification System (USCS) is most commonly used in the construction industry. Complete details on this system can be found in any textbook or manual of soils engineering. (For example, see Soils Manual MS-10, The Asphalt Institute, College Park, Maryland, 1978.)

Soils vary in physical and chemical structure, but can be separated into five broad groups:

Gravel: Individual grains vary from 0.08 to 3 in (2 to 75 mm) in diameter and are generally rounded in appearance.

Sand: Small rock and mineral fragments smaller than 0.08 in (2 mm) in diameter.

Silt: Fine grains appearing soft and floury.

Clay: Very fine textured soil which forms hard lumps when dry and is sticky to slick when wet.

Organic: Peat.

Soils are sometimes classified into categories according to the ability of the soil(s) to enhance the structural performance of the pipe when installed in the particular soil. One such classification is described in ASTM D-2321 "Standard Practice for Underground Installation of Flexible Thermoplastic Sewer Pipe."

The project engineer often requires a soil survey along the route of a proposed pipeline. Information from the survey helps to determine the necessary trench configuration, and to decide if an imported soil will be required to be placed around the pipe. Soil parameters such as soil type, soil density, and moisture content are usually considered in a design. Soil stiffness (modulus) is an extremely important soil property and is the main contributor to the pipe-soil system performance. Experience has shown that a high soil density will ensure a high soil stiffness. Therefore, soil density is usually given special importance in piping system design.

Economy in any design is always a prime consideration. The engineer must consider the cost of compaction as opposed to bringing in a select material such as pea gravel which will flow into place in a fairly dense state. For piping systems, a compacted, well-graded, angular, granular material provides the best structural support. However, such is not always required. In selecting a backfill material the designer will consider such things as depth of cover, depth of water table, pipe materials, compaction methods available, and so forth.

Strength of Materials

There are many types of piping materials on the market today ranging from rigid concrete to flexible thermal plastic. Proponents of each lay claim to certain advantages for their material. Such things as inherent strength, stiffness, corrosion resistance, lightness, flexibility, and ease of joining are some characteristics that are often given as reasons for using a particular material.

A pipe must have enough strength and/or stiffness to perform its intended function. It must also be durable enough to last for its design life. The term "strength" as used here is the ability to resist stress. Stresses in a conduit may be caused by such loadings as internal pressure, soil loads, live loads, differential settlement, and longitudinal bending, to name a few. The term "stiffness" refers to the material's ability to resist deflection. Stiffness is directly related to the modulus of elasticity of the pipe material and the second moment of the cross section of the pipe wall. Durability is a measure of the pipe's ability to withstand environmental effects with time. Such terms as corrosion resistance and abrasion resistance are durability factors.

Piping materials are generally placed in one of two classifications—rigid or flexible. A "flexible" pipe has been defined as one that will deflect at least 2 percent without structural distress. Materials that do not meet this criteria are usually considered to be "rigid." Claims that a particular pipe is neither flexible nor rigid but somewhere in-between have little importance since current design standards are based either on the concept of a flexible conduit or on the concept of a rigid conduit. This important subject will be discussed in detail in subsequent chapters. See Fig. 1.1.

Concrete and clay pipes are examples of materials which are usually considered to be rigid. Thin steel and plastic pipes are usually considered to be flexible. Each type of pipe may have one or more performance limits which must be considered by the design engineer. For

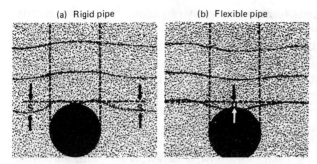

Figure 1.1 The effect of soil settlement on (a) rigid and (b) flexible pipes. S represents settlement of backfill for a rigid pipe. D represents vertical deflection of a flexible pipe as it deflects under earth pressure. (*Reprinted, by permission, from AWWA Manual M11 Steel Pipe Design and Installation, American Water Works Association, 1964.*)

rigid pipes, strength to resist wall stresses due to the combined effects of internal pressure and external load is usually critical. For flexible pipes, stiffness may be important in resisting ring deflection and possible buckling. Each manufacturer or industry goes to great lengths to establish characteristics of their particular product. These parameters are readily available to the design engineer. The desire to have products with high strength has given rise to reinforced products such as steel-reinforced concrete and glass-reinforced thermal setting plastic. For such products, other performance limits often arise such as a strain limit to prevent cracking. For a thermal plastic pipe, such as PVC pipe, strength is measured in terms of a long-term hydrostatic-design hoop stress. Thus, it can be seen that installations of all products will not be designed in exactly the same manner. The engineer must be familiar with design criteria for the various pipe products and know where proper design parameters can be obtained.

Pipe Hydraulics

The field of study of fluid flow in pipes is often referred to as "hydraulics." Designers of water or sewer systems need some knowledge of pipe hydraulics.

Flow in pipes is usually classified as "pressure flow" for systems where pipes are flowing full or "open channel flow" when pipes are not flowing full. Water systems are pressure systems and are considered to be flowing full. On the other hand, sewer systems, for the most part, are open channel systems. The exception to this is for forced sewer mains where lift pumps are used to pump sewage under pressure. The relatively small concentration of solids found in sanitary or storm sewage is not sufficient to make it behave hydraulically significantly different than water. Thus sewage is accepted to have the same hydraulic flow characteristics of water. Of course, the design engineer must be aware of the possibility of the deposition of solid and hydrogen sulfide gas generation in sanitary sewers. These considerations are not within the scope of this text.

In either case, pressure flow or open channel flow, the fluid encounters frictional resistance. This resistance produces head loss which is a function of the inside surface finish or pipe roughness. The smoother the inside surface, the better the flow. Many theories and empirical equations have been developed to describe flow in pipes. The solution of most flow problems requires experimentally derived coefficients which are used in conjunction with empirical equations. For pressure flow, the Hazen-Williams equation is widely accepted. Another equation that has a more theoretical basis is attributed to Darcy and Weisback. For open channel flow, the so-called "Manning equation" is normally used. These equations or others are used to calculate head loss as a function of flow or vice versa.

Water Systems

Water systems are lifelines of communities. They consist of such items as, valves, fittings, thrust restraints, pumps, reservoirs, and of course, pipes and other miscellaneous appurtenances. The water system is sometimes divided into two parts: (1)The transmission lines and (2)the distribution system. The transmission system is that part of the system which brings water from the source to the distribution system. Transmission lines have few if any interconnections. Because of this, flow in such a line is usually considered to be quasisteady with only relatively small transients. Such lines are normally placed in fairly shallow soil cover. The prime design consideration is internal pressure. Other design considerations include longitudinal stresses, ring deflection, buckling, and thrust restraints.

The distribution piping system distributes water to the various users. It includes many connections, loops, and so forth. The design is somewhat similar to that of transmission lines except a substantial surge allowance for possible water hammer is included in the pressure design. Also, more care is usually taken in designing the backfill for around the pipe, fittings, and connections. This is done to prevent longitudinal bending and differential settlement. Distribution systems are made up of an interconnected pipe network. The hydraulic analysis of such a system is almost impossible by "hand" methods, but is readily accomplished using programming methods via digital computers.

The piping system must be strong enough to withstand induced stresses, have relatively smooth walls, have a tight joining system, and be somewhat chemically inert with respect to soil and water. Normal design life for such a system should be 50 years minimum.

Wastewater Systems

A sewage system is made up of a collection system and a treatment system. We are concerned only with the collection part. For the most part, sanitary sewers and storm (street) sewers are separate. However, there are a few older cities in North America which use "combined sewers." The ills of these combined sewers have been recognized by modern engineers and such systems are no longer designed. Most state and regional engineering and public works officials and agencies no longer permit installation of these dual purpose lines. Unfortunately, some combined sewers still exist in the United States.

Some sanitary sewers are pressurized lines (sewer force mains), but most are gravity flow lines. The sanitary sewer is usually buried quite deep to allow for the pickup of water flow from basements. Due to this added depth, higher soil pressures which act on the pipe, are probable. To resist these pressures, pipe strength and/or pipe stiffness become important parameters in the design. Soil backfill, its placement and compaction also become important to the design engineer. The installation may take place below the water table so construction procedures may include dewatering and wide trenching. For such a system, the pipe should be easy to join with a tight joint that will prevent infiltration. The soil-pipe system should be designed and constructed to support the soil load. The pipe material should be chemically inert with respect to soil and sewage, including possible hydrogen sulfides. The inside wall should be relatively smooth so as not to impede the fluid flow.

Storm sewer design conditions are not as rigorous as they are for sanitary sewers. Storm sewers are normally not as deep. The requirements for the joining system are often very lax and usually allow exfiltration and infiltration. Because of the above, loose joining systems are often acceptable for storm sewers. The design life for any sewer system should also be 50 years minimum.

Design for Value

Most piping system contracts are awarded to the lowest bidder. Contractors will usually bid materials and construction methods which allow for the lower initial cost with little thought to future maintenance or life of the system. Even for the owner, the lowest initial cost is often the overriding factor. However, the owner and the engineer should insist on a design based on value. For engineers, economics is always an important consideration; and any economic evaluation must include more than just initial cost. Annual maintenance and life of the system should also be considered.

Initial cost may include such things as piping materials, trenching, select backfill, compaction, site improvements and restoration, and engineering and inspection. Pipe cost is related to pipe material and to pipe diameter. Diameter is controlled by the design flow rate and pipe roughness. That is, a smaller diameter may be possible if a pipe with a smooth interior wall is selected. Annual maintenance cost includes cleaning, repair and replacement due to erosion, corrosion, and so forth. Life is directly related to durability and is affected by such things as corrosion, erosion, and other types of environmental degradation.

Often, the question is not whether the pipe will last, but how long it will perform its designed function. Generally, metals corrode in wet clay soils and corrode at an accelerated rate in the presence of hydrogen sulfide sewer gas. Concrete-type structures are also attacked by hydrogen sulfide and the resulting sulfuric acid. Care should be taken when selecting a pipe product for any service application to ensure that environmental effects upon the life of the system have been taken into consideration. The system should be designed for value.

External Loads

Loads are exerted on buried pipes by the soil that surrounds them. Methods for calculating these loads are given in this chapter. Marston's Theory for loads on buried conduits is discussed along with the various factors which contribute to these loads. Underground pipes are placed in tunnels, buried under highways, buried under railways, and buried under airports. Methods are given for the determination of loads which are imposed on pipes in these and other applications.

Soil Pressure

The horseless carriage had its volume-production start with the Oldsmobile in 1902 and the need for improved roads was immediately apparent. Many projects for road drainage were begun using clay tile and concrete drain tile. One major problem existed, however. There was no rational method of determining the earth load these buried drains would be subjected to. As a result, there were many failures of pipelines.

The loads imposed on conduits buried in the soil depend upon the stiffness properties of both the pipe structure and the surrounding soil. This results in a statically indeterminant problem in which the pressure of the soil on the structure produces deflections that, in turn, determine the soil pressure. The subject of soil structure interaction has been of engineering interest since the early 1900s.

Rigid pipe

Marston load theory.   Anson Marston, who was dean of engineering at Iowa State University, investigated the problem of determining loads on buried conduits. In 1913, Marston published his original paper entitled "The Theory of Loads on Pipes in Ditches and Tests of Cement and Clay Drain Tile and Sewer Pipe."[4] This work was the beginning of methods for calculating earth loads on buried pipes. The formula is now recognized the world over as the Marston load equation. More recently, demands to protect and improve our environment, and rising construction costs have produced research that has substantially increased our knowledge of soil-structure interaction phenomenon. However, much of this knowledge has yet to be applied to design practice. Many questions are as yet unresolved.

Trench condition.   The Marston load theory is based on the concept of a prism of soil in the trench that imposes a load on the pipe as shown in Fig. 2.1. A trench (ditch) conduit as defined by Marston was a relatively narrow ditch dug in undisturbed soil. Marston reasoned that settlement of the backfill and pipe generates shearing or friction forces at the sides of the trench. He also assumed that cohesion would be negligible since (1) considerable time would have to elapse before cohesion could develop and (2) the assumption of no cohesion would yield the maximum load on the pipe.

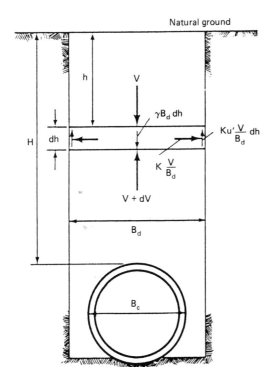

Figure 2.1 Basis for Marston's theory of loads on buried pipe. (W_d = load on conduit per unit length along conduit in pounds per linear foot; e = base of natural logarithms; γ = unit weight of backfill, i.e., pounds per cubic foot; V = vertical pressure on any horizontal plane in backfill, in pounds per unit length of ditch; B_c = horizontal breadth (outside) of conduit, in feet; B_d = horizontal width of ditch at top of conduit, in feet; H = height of fill above top of conduit, in feet; h = distance from ground surface down to any horizontal plane in backfill, in feet; C_d = load coefficient for ditch conduits; μ = tan ϕ = coefficient of internal friction of backfill; μ' = tan ϕ = coefficient of friction between backfill and sides of ditch; and K = ratio of active lateral unit pressure to vertical unit pressure.) *(Reprinted from Soil Engineering, Merlin G. Spangler and Richard L. Handy, Harper & Row, 1973, by permission of the publisher.)*

The vertical pressure, V at the top of any differential volume element $[B_d(1)]dh$ is balanced by an upward vertical force at the bottom of the element $V + dV$ (see Fig. 2.1). The volume element is B_d wide, dh in height, and of unit length along the axis of the pipe and trench. The weight of the elemental section is its volume times its unit weight expressed as

$$w = B_d(dh)\,(1)\gamma$$

where $B_d\,(dh)\,(1)$ is volume of the element and γ is the specific weight density.

The lateral pressure, P_L, at the sides of the element at depth h is

$$P_L = \frac{\text{active lateral unit pressure}}{\text{vertical unit pressure}} \times \text{vertical unit pressure}$$

or

$$P_L = K \text{ (Rankine's ratio)} \times \left(\frac{V}{B_d}\right)$$

The shearing forces per unit length, F_s, on the sides of the differential element, induced by these lateral pressures, are $F_s = K(V/B_d)\,(\mu')dh$, where $\mu' = $ coefficient of friction. The vertical forces on the element are summed and set equal to zero.

$$F_v = 0$$

Or, the upward vertical forces are equal to the downward vertical forces. Thus, for equilibrium:

Vertical force at bottom + shear force at sides = vertical force at top
+ weight of the element

$$(V + dV) + \left(\frac{2K\mu'V}{B_d}\right)dh = V + \gamma B_d dh$$

(dimensionally force per length)

$$0 = \left(B_d - \frac{2K\mu'V}{B_d}\right)\left(\frac{dh}{dV}\right) \tag{2.1}$$

The solution to the differential equation [Eq. (2.1)] is

$$V = \frac{\gamma B_d^2[1 - e^{-2K\mu'(h/B_d)}]}{2K\mu'} \qquad (2.2)$$

Substituting $h = H$, we get the total vertical pressure at the elevation of the top of the conduit. How much of this vertical load V is imposed on the conduit is dependent upon the relative compressibility (stiffness) of the pipe and soil. For very rigid pipe (clay, concrete, heavy-walled cast iron, and so forth), the sidefills may be very compressible in relation to the pipe and the pipe may carry practically all of the load V. For flexible pipe, the pipe may be less rigid than the sidefill soil. The maximum load on ditch conduits is expressed in Eq. (2.2) with $h = H$. For simplicity and ease of calculation, the load coefficient C_d is defined

$$C_d = \frac{1 - e^{-2K\mu'(H/B_d)}}{2K} \qquad (2.3)$$

Now the load on a rigid conduit in a ditch is expressed as

$$W_d = C_d \gamma B_d^2 \qquad (2.4)$$

The function

$$C_d = \frac{1 - e^{-2K\mu'(H/B_d)}}{2K}$$

is then plotted as H/B_d versus C_d, for various soil types as defined by their $K\mu'$ values where $K\mu'$ is a function of the coefficient of internal friction of the fill material (see Fig. 2.2).

The values of K, μ, and μ' were determined experimentally by Marston and typical values are given in Table 2.1.

Example Problem 2.1   What is the maximum load on a very rigid pipe in a ditch excavated in sand? The pipe diameter (OD) is 18 in, the trench width is 42 in, the depth of burial is 8 ft and the soil unit weight is 120 lb/ft³.

TABLE 2.1   Approximate Values of Soil Unit Weight, Ratio of Lateral to Vertical Earth Pressure, and Coefficient of Friction against Sides of Trench

Soil type	Unit weight, lb/ft³	Rankine's ratio K	Coefficient of friction μ
Partially compacted damp top soil	90	0.33	0.50
Saturated top soil	110	0.37	0.40
Partially compacted damp clay	100	0.33	0.40
Saturated clay	120	0.37	0.30
Dry sand	100	0.33	0.50
Wet sand	120	0.33	0.50

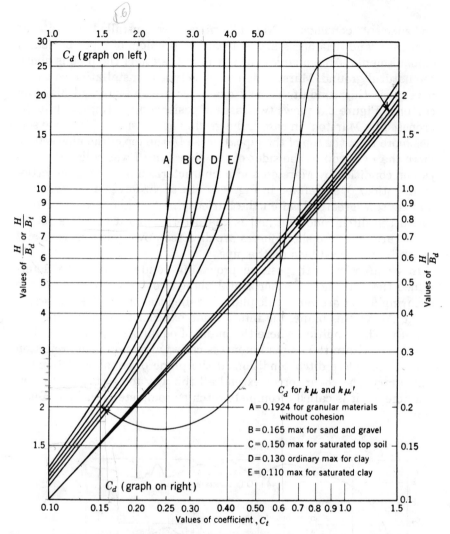

Figure 2.2 Computation diagram for earth loads on trench conduits completely buried in trenches. *(Reprinted, by permission, from Design & Construction of Sanitary & Storm Sewers, "Manuals & Reports on Engineering Practice No. 37," American Society of Civil Engineers and "Manual of Practice No. 9," Water Pollution Control Federation, 1969, p. 189.)*

1. Determine C_d:

   From Table 2.1 for sand, $K\mu = K\mu' = 0.165$

   From Fig. 2.2, $C_d = 1.6$

   $$\frac{H}{B_d} = \frac{8 \text{ ft}}{42 \text{ in}} \times \frac{1 \text{ ft}}{12 \text{ in}} = 2.29$$

2. Calculate load from Eq. (2.4):

   $$W_d = C_d \gamma B_d{}^2 = 1.6(120)\left(\frac{42}{12}\right)^2 = 2352 \text{ lb/ft}$$

Embankment conditions. Not all pipes are installed in ditches (trenches) and it is necessary to treat the problem of pipes buried in embankments. An embankment is where the top of the pipe is above the natural ground. Marston defined this type of installation as a positive projecting conduit. Typical examples are railway and highway culverts. Figure 2.3 shows two cases of positive projecting conduits as proposed by Marston. In case I, the ground at the sides of the pipe settles more than the top of the pipe. In case II, the top of the pipe settles more than the soil at the sides of the pipe. Case I was called the projection condition by Marston and is characterized by a positive settlement ratio, r_{sd}, as defined in Fig. 2.4. The shear forces are downward and cause a greater load on the buried pipe for this case. Case II is called the ditch condition and is characterized by a negative settlement ratio, r_{sd}. The shear forces are directed upward in this case and result in a reduced load on the pipe.

In conjunction with positive projecting conduits, Marston determined the existence of a horizontal plane above the pipe, where the shearing forces are zero. This plane is called the plane of equal settlement. Above this plane, the interior and exterior prisms of soil settle equally. The condition where the plane of equal settlement is real (it is located within the embankment) is called an incomplete projection or an incomplete ditch condition. If the plane of equal settlement is imaginary (the shear forces extend all the way to the top of the embankment), it is called a complete ditch or complete projection condition.

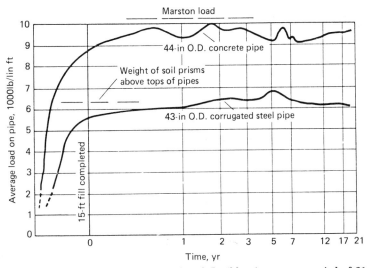

Figure 2.3 Measured loads on rigid and flexible pipe over a period of 21 years. *(Reprinted from Spangler and Handy, Soil Engineering, Harper & Row, 1973, by permission of the publisher.)*

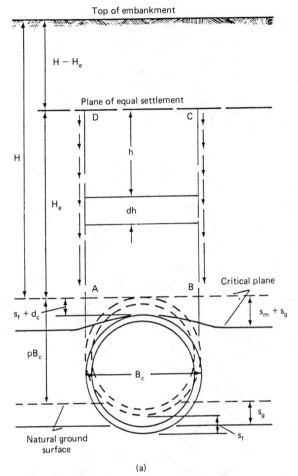

Figure 2.4 Comparison of positive projecting conduits: (*a*) Projection condition; (*b*) ditch condition. ($r_{sc} = [(S_m + S_g) - (S_f + d_c)]/S_m$; r_{sd} = settlement ratio; s_m = compression of soil at sides of pipe; s_g = settlement of natural ground surface at sides of pipe; s_f = settlement of foundation underneath pipe; d_c = deflection of the top of pipe.) (*Reprinted from Spangler and Handy, Soil Engineering, Harper & Row, 1973, by permission of the publisher.*)

All of the above discussed parameters affect the load on the pipe and are incorporated in Marston's load equation for positive projecting (embankment) conduits.

$$W_c = C_c \gamma B_c^{\,2} \qquad\qquad (2.5)$$

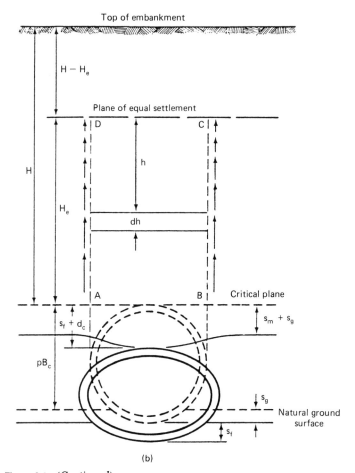

Figure 2.4   (*Continued*).

where

$$C_c = \frac{e^{\pm 2K\mu(H/B_c)} - 1}{\pm 2K\mu} \tag{2.6}$$

or

$$C_c = \frac{e^{\pm 2K\mu(H_e/B_c)} - 1}{\pm 2K\mu} + \left(\frac{H}{B_c} - \frac{H_e}{B_c}\right)e^{\pm 2K\mu(H_e/B_c)} \tag{2.7}$$

Equation (2.6) is for the complete condition. The negative signs are for the complete ditch and the positive signs are for the complete projection condition.

Equation (2.7) is for the incomplete condition, where the negative signs are for the incomplete ditch and the positive signs are for the incomplete projection condition. H_e is the height of the plane of equal settlement. Note that if $H_e = H$, the incomplete case [Eq. (2.7)] becomes the complete case and Eq. (2.6) applies for C_c.

Although the above equations are difficult and cumbersome, they have been simplified and can be found in graphical form in many references.

Note that value C_c is a function of the ratio of height of cover to pipe-diameter ratio, the product of the settlement ratio and projection ratio, Rankine's constant and the coefficient of friction.

$$C_c = f\left(\frac{H}{B_c}, r_{sd}p, K, \mu\right)$$

The value of the product $K\mu$ is generally taken as 0.19 for the projection condition and 0.13 for the ditch condition. Figure 2.5 is a typ-

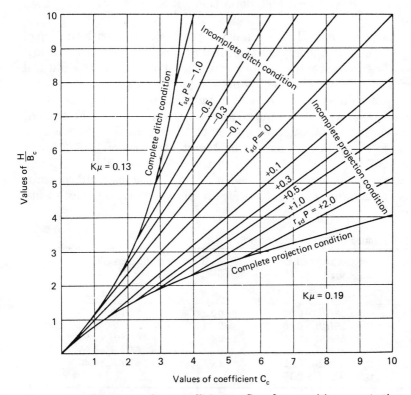

Figure 2.5 Diagram for coefficient C_c for positive projecting conduits. *(Reprinted from Spangler and Handy, Soil Engineering, Harper & Row, 1973,[11] by permission of the publisher.)*

ical diagram of C_c for the various values of H/B_c and $r_{sd}p$ encountered. Table 2.2 gives the equations of C_c as a function of H/B_c for various values of $r_{sd}p$ and $K\mu$.

The settlement ratio r_{sd} is difficult, if not impossible, to determine even empirically from direct observations. Experience has shown that the values tabulated in Table 2.3 can be used with success. Note that when $r_{sd}p = 0$, $C_c = H/B_c$ and $W_c = H\gamma B_c$. This is the prism load (that is, the weight of the prism of soil over the top of the pipe). When $r_{sd} = 0$, the plane at the top of the pipe called the critical plane settles the same amount as the top of the conduit (see Fig. 2.4). The settlement ratio is defined as

$$r_{sd} = \frac{(S_m + S_g) - (S_f + d_c)}{S_m} \tag{2.8}$$

Critical plane settlement = S_m (strain in side soil) + S_g (ground settlement). Settlement of the top of the pipe = S_f (conduit settlement) + d_c (vertical pipe deflection). If $S_m + S_g = S_f + d_c$, then $r_{sd} = 0$.

When a pipe is installed in a narrow shallow trench with the top of the pipe level with the adjacent natural ground, the projection ratio p is zero. The distance from the top of the structure to the natural ground surface is represented by pB_c.

The question may be asked, is Marston's equation for the earth load on a rigid pipe in a ditch valid regardless of the width of the trench? The answer to this question was given by W. J. Schlick, a colleague of Marston in 1932.[9] Schlick found that Marston's equation [Eq. (2.4)] for W_d was valid until the point where the ditch conduit load W_d was equal to the projection conduit load W_c. That is, the load will continue to increase according to Eq. (2.4) for an increasing trench width until

TABLE 2.2   Values of C_c in Terms of H/B_c

Incomplete projection condition $K\mu = 0.19$		Incomplete ditch condition $K\mu = 0.13$	
$r_{sd}p$	Equation	$r_{sd}p$	Equation
+ 0.1	$C_c = 1.23H/B_c - 0.02$	− 0.1	$C_c = 0.82H/B_c + 0.05$
+ 0.3	$C_c = 1.39H/B_c - 0.05$	− 0.3	$C_c = 0.69H/B_c + 0.11$
+ 0.5	$C_c = 1.50H/B_c - 0.07$	− 0.5	$C_c = 0.61H/B_c + 0.20$
+ 0.7	$C_c = 1.59H/B_c - 0.09$	− 0.7	$C_c = 0.55H/B_c + 0.25$
+ 1.0	$C_c = 1.69H/B_c - 0.12$	− 1.0	$C_c = 0.47H/B_c + 0.40$
+ 2.0	$C_c = 1.93H/B_c - 0.17$		

SOURCE: From Ref. (1). Reprinted from *Soil Engineering*, 3d ed., Merlin G. Spangler and Richard L. Handy, Harper & Row, 1973. By permission of the publisher.

TABLE 2.3   Design Values of Settlement Ratio

Conditions	Settlement ratio
Rigid culvert on foundation of rock or unyielding soil	+1.0
Rigid culvert on foundation of ordinary soil	+0.5 to +0.8
Rigid culvert on foundation of material that yields with respect to adjacent natural ground	0 to +0.5
Flexible culvert with poorly compacted side fills	−0.4 to 0
Flexible culvert with well-compacted side fills*	−0.2 to +0.8

*Not well established.
SOURCE: Reprinted from *Soil Engineering*, 3d ed., Merlin G. Spangler and Richard L. Handy, Harper & Row, 1973. By permission of the publisher.

the ditch load is equal to the embankment load. Once this point is reached, the correct load will be calculated by Eq. (2.5). The trench width at which this occurs is called the transition width. Figure 2.6 is a plot of values of H/B_c and $R_{sd}p$ that give B_d/B_c values that represent the transition width. That is, $W_c = W_d$. It is generally suggested that an $r_{sd}p$ value of 0.5 be used to determine the transition width.

If the calculation of B_d/B_c is

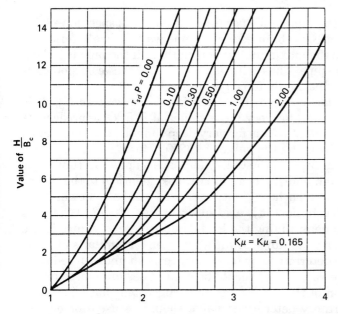

Values of $\dfrac{B_d}{B_c}$ for ditch conduit leads to equal project conduit loads

Figure 2.6 Curves for transition-width ratio. *(Reprinted from Spangler and Handy, Soil Engineering, Harper & Row, 1973,[11] by permission of the publisher.)*

Greater than that of Fig. 2.6, use W_d

Less than that of Fig. 2.6, use W_c

Equal to that of Fig. 2.6, then $W_c = W_d$

Example Problem 2.2   What is the transition width for a 12 in pipe buried 6 ft deep?

$$H/B_c = \frac{6 \text{ ft}}{12 \text{ in}} \times \frac{12 \text{ in}}{1 \text{ ft}} = 6$$

From Fig. 2.6,

$$B_d/B_c = 2.35$$

$$r_{sd}p = 0.5$$

$$B_d(\text{transition}) = \frac{Bd}{B_c}B_c = \frac{2.35(12)}{12} = 2.35 \text{ ft}$$

Tunnel construction.   Marston's theory may be used to determine soil loads on pipes that are in tunnels or that are jacked into place through undisturbed soil. The Marston tunnel-load equation is as follows:

$$W_t = C_tB_t(\gamma B_t - 2C) \tag{2.9}$$

W_t is the load on the pipe in lbs per linear ft, and γ is specific weight. The load coefficient C_t is obtained in the same way C_d was determined (see Fig. 2.2). B_t is the maximum tunnel width, or if the pipe is jacked, B_t is the OD of the pipe. The coefficient C is called the "cohesion coefficient" and is dimensionally force per unit area (lb/ft^2).

Equation (2.3) can be used in calculating C_t as well as C_d. This equation indicates that, for very large values of H/B, C_t approaches a limiting value of $1/(2K\mu')$. Thus for very deep tunnels, the load can be closely estimated by using the value of $1/(2K\mu')$ for C_t.

It is readily apparent that the theory for loads on pipes in tunnels or being jacked through undisturbed soil is almost identical to the theory for loads on pipes in trenches. The tunnel load will be somewhat less because of the soil cohesion. It is also apparent from Eq. (2.9) that C is very important in determining the load. Unfortunately, values of the coefficient C have a wide range of variation even for similar soils. The value of C may be determined by laboratory tests on undisturbed samples. Conservative values of C should be used in design to account for possible saturation of the soil. It has been suggested that about one-third of the laboratory determined value should be used for design. The Water Pollution Control Federation (WPCF) Manual of Practice No. FD-5 recommends the use of values given in Table 2.4 if reliable laboratory data are not available or if such tests are impractical. It is also suggested that this coefficient be taken as zero for any zone subjected to seasonal frost and cracking or loss of strength because of saturation. The factor $\gamma B_t - 2C$ cannot be negative. Therefore, $2C$ cannot be larger than γB_t.

TABLE 2.4   Recommended Safe Values of Cohesion, C

Material	Values of C	
	kPa	lb/ft^2
Clay, very soft	2	40
Clay, medium	12	250
Clay, hard	50	1000
Sand, loose dry	0	0
Sand, silty	5	100
Sand, dense	15	300

Flexible pipe

A flexible pipe derives its soil-load carrying capacity from its flexibility. Under soil load, the pipe tends to deflect, thereby developing passive soil support at the sides of the pipe. At the same time, the ring deflection relieves the pipe of the major portion of the vertical soil load which is picked up by the surrounding soil in an arching action over the pipe. The effective strength of the flexible pipe-soil system is remarkably high. For example, tests at Utah State University indicate that a rigid pipe with a three-edge bearing strength of 3300 lb/ft buried in class C bedding will fail by wall fracture with a soil load of about 5000 lb/ft. However, under identical soil conditions and loading, a PVC sewer pipe deflects only 5 percent. This is far below the deflection that would cause damage to the PVC pipe wall. Thus the rigid pipe has failed, but the flexible pipe performed successfully and still has a factor of safety with respect to failure of 4 or greater. Of course, in flat plate or three-edge loading, the rigid pipe will support much more than the flexible pipe. This anomaly tends to mislead some engineers because they relate low flat-plate supporting strength with in-soil load capacity—something one can do for rigid pipes but cannot do for flexible pipes.

Marston load theory.   For the special case when the sidefill and pipe have the same stiffness, the amount of the load V that is proportioned to the pipe can be done merely on a width basis. This means that if the pipe and the soil at the sides of the pipe have the same stiffness, the load V will be uniformly distributed as shown in Fig. 2.7. By simple proportion the load becomes

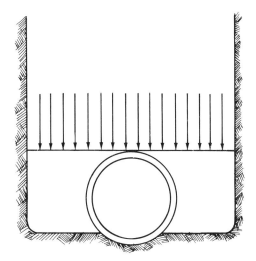

Figure 2.7 Load proportioning according to Marston's theory for a flexible pipe.

$$W_c = \frac{W_d B_c}{B_d} = \frac{C_d \gamma B_d^2 B_c}{B_d} \cdot W_c = C_d \gamma B_c B_d \qquad (2.10)$$

Pipe stiffness versus soil compressibility. Measurements made by Marston and Spangler revealed that the load on a flexible pipe is substantially less than on a rigid pipe (see Fig. 2.3). The magnitude of this difference in loads may be a little shocking. The following analogy will help to understand what happens in the ground as a flexible pipe deflects. Suppose a weight is placed on a spring. We realize the spring will deform resisting deflection because of its spring stiffness. When load versus deflection is plotted, we find that this relationship is linear up to the proportional limit of the spring. When a load is placed on a flexible pipe, it also deflects but resists deflection because of its stiffness. It is even possible to think of soil as being a nonlinear spring that resists movement or deflection because of its stiffness.

When we draw an analogy between a rigid pipe represented by a stiff spring in comparison to soil at its sides, represented by more flexible springs, and then place a load or weight on this spring system representing a rigid pipe in soil, we can easily visualize the soil deforming and the pipe carrying the majority of the load. If the situation is reversed and we place a flexible spring between two springs which are much stiffer representing the soil, we can again picture the pipe deflecting as a load is applied and the soil in this case being forced to carry the load to a greater extent.

Equation (2.10) has become known as the Marston load equation for flexible pipes. It should be remembered, however, that the assumption of the soil and the pipe stiffness being equal has been used in its development and that it should not be used merely because a pipe is flexible.

The maximum load as predicted by the Marston equations [Eqs. (2.4) and (2.10)] does not take place instantaneously and may not occur for some time. In certain cases the initial load may be 20 to 25 percent less than the maximum load predicted by Marston.

Example Problem 2.3 For Example Problem 1, what would be the load if the pipe and side soil had approximately the same stiffness?

$$W_c = C_d\gamma B_c B_d = 1.6\,(120)\left(\frac{42}{12}\right)\left(\frac{18}{12}\right) = 1008 \text{ lb/ft} \qquad (2.10)$$

Prism load. It should be pointed out that Eq. (2.4) represents a maximum-type loading condition and Eq. (2.10) represents a minimum. For a flexible pipe, the maximum load is always much too large since this is the load acting on a rigid pipe. The minimum is just that, a minimum. The actual load will lie somewhere between these limits.

A more realistic design load for a flexible pipe, would be the prism load which is the weight of a vertical prism of soil over the pipe. Also, a true trench condition may or may not result in significant load reductions on the flexible conduit since a reduction depends upon the direction of the frictional forces in the soil. Research data indicate that the effective load on a flexible conduit lies somewhere between the minimum predicted by Marston and the prism load. On a long-term basis, the load may approach the prism load. Thus, if one desires to calculate the effective load on a flexible conduit, the prism load is suggested as a basis for design. The prism or embankment load is given by the following equation (see Fig. 2.8):

$$P = \gamma H \qquad (2.11)$$

where P = pressure due to weight of soil at depth H
γ = unit weight of soil
H = depth at which soil pressure is required

Example Problem 2.4 Assume an 8-in OD flexible pipe is to be installed in a 24-in-wide trench with 10 ft of clay soil cover. The unit weight of the soil is 120 lb/ft^3. What is the load on the pipe?

Marston load: Use Eq. (2.10) for minimum W.
$$W_d = C_d\gamma B_c B_d$$

where C_d = 2.8 from Fig. 2.2
γ = 120 lb/ft^3
B_d = trench width = 2 ft
B_c = OD = 8 in = ⅔ ft
Marston load = W_d = (2.8)(120)(2)(⅔) = 448 lb/ft

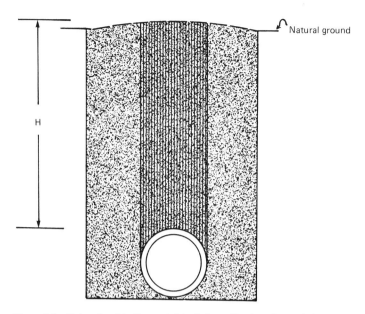

Figure 2.8 Prism load is the weight of the soil prism located directly above the pipe.

Prism load: Use Eq. (2.11)..

$$P = \gamma H = 120(10) = 1200 \text{ lb/ft}^2$$

To obtain load in lb/ft units, multiply above by the pipe OD in feet.

$$W = 1200(\tfrac{2}{3}) = 800 \text{ lb/ft}$$

The Marston load for this example is 56 percent of the prism load and is unconservative for design. Again, for flexible conduits, the prism-load theory represents a realistic estimate of the maximum load and is slightly conservative.

Trench condition. The Marston-Spangler equation for the load on a flexible pipe in a trench is given by Eq. (2.10). The load coefficient C_d is obtained from Fig. 2.2. One may ask, under what conditions, if any, will the prism load and the ditch (trench) load be equal.

$$\text{Prism load} = P = \gamma H \qquad \text{lb/ft}^2$$

$$\text{Marston load} = W_d = C_d \gamma B_d B_d$$

Multiply the prism load by B_c (to express in lb/ft as in Marston load) and set equal to the Marston load.

$$PB_c = \gamma H B_c = W_d = C_d \gamma B_d B_c$$

Solve for

$$C_d = \frac{H}{B_d}$$

Thus the prism load is a special case of the Marston-Spangler trench load. $C_d = H/B_d$ is plotted as a straight 45-degree line in Fig. 2.2. One of the advantages of the prism load is that it is independent of trench width.

Embankment condition. The load on a flexible pipe in an embankment may be calculated by the Marston-Spangler theory via Eq. (2.5).

$$W_c = C_c \gamma B_c^2$$

This equation does not include a trench width term since a trench is not involved. Again it is interesting to set this load equal to the prism load.

$$\text{Prism load} \times B_c = PB_c = \gamma H B_c$$

$$\text{Marston embankment load } W_c = C_c \gamma B_c^2$$

Setting equal

$$\gamma H B_c = C_c \gamma B_c^2$$

or

$$C_c = H/B_c$$

C_c can be determined from Fig. 2.5. The above equation plots as a straight 45-degree line on Fig. 2.5. This is the line shown for $r_{sd}P = 0$. Thus for an embankment, the prism load is the same as the Marston load for $r_{sd}P = 0$.

Tunnel loadings. There is little documented data dealing with loads on flexible pipes placed in unsupported tunnels. However, since a flexible pipe develops a large percentage of its load carrying capacity from passive side support, this support must be provided or the pipe will tend to deflect until the sides of the pipe are being supported by the sides of the tunnel.

When a flexible pipe is jacked into undisturbed soil, the load may be calculated by either the prism load, Eq. (2.11), or by Eq. (2.9).

$$B_t = B_c$$

$$W_p = PB_t = \gamma H B_t \qquad (2.11)$$

$$W_t = C_t B_t (\gamma B_t - 2C) \qquad (2.9)$$

The prism load in this case will be very conservative because it neglects not only friction but also the cohesion of the soil. If C_t is taken as H/B_t and the cohesion coefficient is zero, then the two methods of calculating loads give the same results.

Longitudinal Loading

Certain types of pipe failures which have been observed over the years, are indicative of the fact that only under ideal conditions is a pipeline truly subjected to only vertical earth loading. There are other forces that in some way produce axial bending stresses in the pipe. These forces can be large, highly variable, localized, and may not lend themselves to quantitative analysis with any degree of confidence. Some of the major causes of axial bending or beam action in a pipeline area are

1. Nonuniform bedding support
2. Differential settlement
3. Ground movement for such external forces as earthquakes or frost heave

Nonuniform bedding support

A nonuniform bedding can result from unstable foundation materials, uneven settlement due to overexcavation and nonuniform compaction, and undermining, such as might be produced by erosion of the soil into a water course or by a leaky sewer.

One of the advantages of a flexible conduit is its ability to deform and move away from pressure concentrations. The use of flexible joints also enhances a pipe's ability to yield to these forces and reduce the risk of rupture. These advantages, coupled with good engineering and a proper installation, virtually eliminate axial bending as a cause of failure in a flexible pipe. The examples which follow in Figs. 2.9, 2.10, and 2.11 give an indication of the magnitude of bending movements that might be induced.

Axial bending of a long tube in a horizontal plane will produce vertical ring deflection ($\Delta y/D$) due to the bending moments created. Reissner[8] has amplified the work of others in this area and the following formula results from his work on pure bending of a long pressurized tube.

$$\frac{\Delta y}{D} = \frac{1}{16} \left(\frac{D}{t}\right)^2 \left(\frac{D}{R}\right)^2 \qquad (2.12)$$

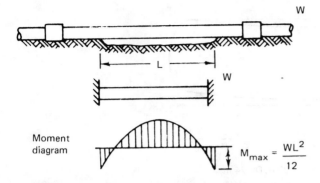

Figure 2.9 Longitudinal bending of conduits.

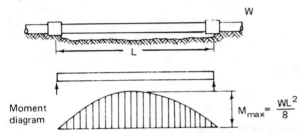

Figure 2.10 Longitudinal bending of conduits.

where D = nominal pipe diameter
$\quad\quad t$ = pipe thickness
$\quad\quad R$ = radius of curvature of the longitudinally deflected pipe
$\quad\Delta y/D$ = ring deflection

Although Reissner's derivation included internal pressure, it has been omitted from Eq. (2.12) because the nonpressure case is the more critical for ring deflection (see Fig. 2.12). This type of bending frequently occurs when pipes are bent around corners.

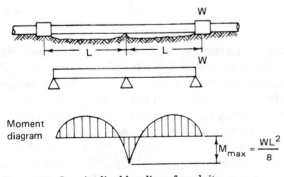

Figure 2.11 Longitudinal bending of conduits.

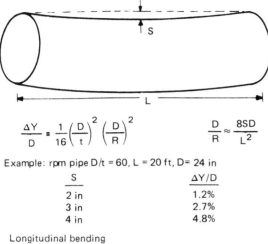

$$\frac{\Delta Y}{D} = \frac{1}{16}\left(\frac{D}{t}\right)^2 \left(\frac{D}{R}\right)^2 \qquad\qquad \frac{D}{R} \approx \frac{8SD}{L^2}$$

Example: rpm pipe D/t = 60, L = 20 ft, D= 24 in

S	$\Delta Y/D$
2 in	1.2%
3 in	2.7%
4 in	4.8%

Longitudinal bending

Buckling radius of curvature R_b

$$R_b = \frac{D}{1.12\ t/D}$$

Figure 2.12   Ring deflection due to axial bending.

Differential settlement

Differential settlement of a manhole or other structure to which the pipe is rigidly connected can induce not only high bending moments but also shearing forces. These forces and moments are set up when the structure and/or the pipe move laterally with respect to the other. Quantitatively, these induced stresses are not easily evaluated. Effort should be taken during design and during construction to see that differential settlement is eliminated or at least minimized. This can be accomplished by the proper preparation and compaction of foundation and bedding materials for both the structure and the connecting pipe.

Ground movement

Variable moisture conditions.   Certain types of soils (mostly expansive clays) are influenced by moisture content. Such soil may be subjected to seasonal rise and fall due to changes in moisture. Good practice does not allow pipes to be bedded directly in such soils. Nevertheless, such shifting by adjacent soil can and will affect a pipeline. Normally these movements are relatively small but may be large enough to adversely affect the pipe performance. To mitigate such adverse effects for rigid pipe, short lengths are used with flexible joints. In the case of flexible pipe, the pipe's natural flexibility tends to allow the pipe to conform to these movements without structural distress. In this case, both longitudinal flexibility and diametrical flexibility are important.

Tidal water may also cause ground movement. These movements may also be designed for, as described above.

Earthquake. In certain critical zones, large ground movement associated with an earthquake may be devastating to a pipeline. These critical zones are primarily those where high differential movement takes place such as a fault zone, a soil shear plane, or transition zones where the pipe enters a structure. Also certain soils will tend to liquify during the earthquake vibration and buried pipelines may rise or tend to float. On the other hand, most buried flexible pipelines can survive an earthquake. Again, a more flexible piping material with a flexible joint will allow the pipe to conform to the ground movement without failure.

Wheel Loading (Live Loads)

Boussinesq solution

Here, live loads mean static or quasistatic surface loads. Buried conduits may be subjected to such applied loads produced by ground transportation traffic. The French mathematician, Boussinesq, calculated the distribution of stresses in a semi-infinite elastic medium due to a point load applied at its surface. This solution assumes an elastic, homogeneous, isotropic medium which soil certainly is not. However, experiments have shown that the classical Boussinesq solution, when properly applied, gives reasonably good results for soil.

Figure 2.13 compares the percent of a surface load that is felt by a buried pipe as a function of depth of burial as calculated by the Boussinesq equation and as found from measurements.

Hall and Newmark integrated the Boussinesq solution to obtain load coefficients. The integration developed by Hall for C_s is used for calculating concentrated loads (such as a truck wheel) and is given in the following form:

$$W_{sc} = C_s\left(\frac{PF'}{L}\right) \tag{2.13}$$

where W_{sc} = load on pipe, lb/unit length
P = concentrated loads, lb
F' = impact factor (see Table 2.5)
L = effective length of conduit (3 ft or less), ft
C_s = load coefficient which is function of $B_c/(2H)$ and $L/(2H)$, where H = height of fill from top of pipe to ground surface, ft; and B_c = diameter of pipe, ft

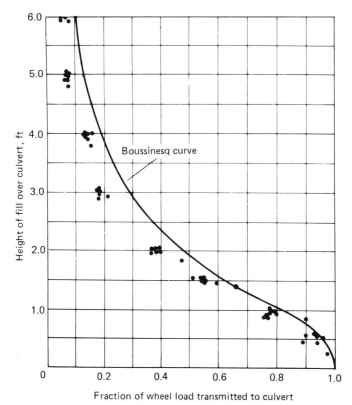

Figure 2.13 Distribution of surface live loads versus loads on a plane at depths of cover. Bouussinesq solutions versus actual measurement. *(Reprinted from Spangler and Handy, Soil Engineering, Harper & Row, 1973,*[11] *by permission of the publisher.)*

TABLE 2.5 Impact Factor (F) versus Height of Cover

	Installation Surface Condition			
Height of cover, ft	Highways	Railways	Runways	Taxiways, aprons, hardstands, run-up pads
0 to 1	1.50	1.75	1.00	1.50
1 to 2	1.35	*	1.00	†
2 to 3	1.15	*	1.00	†
Over 3'	1.00	*	1.00	†

*Refer to data available from American Railway Engineering Association (AREA).
†Refer to data available from Federal Aviation Administration (FAA).
SOURCE: Reprinted from *Uni-Bell Handbook* by permission.

The integration developed by Newmark for C_s is used for calculating distributed loads and is given in the following form

$$W_{sd} = C_s p F' B_c \qquad (2.14)$$

where the only new term is p, which is the intensity of the distributed load, lb/ft^2.

C_s, the load coefficient, is a function of $D/(2H)$ and $M/(2H)$, where D and M are the width and length, respectively, of the area over which the distributed load acts.

The values of the impact factor F' can be determined from Table 2.5 and the load coefficient C_s from Table 2.6.

Highway and railway loads

Figure 2.14 is a plot of an H20 live load, prism earth load, and the sum of the two. An H20 loading is designed to simulate a highway load of a 20-ton truck. Figure 2.14 includes a 50 percent impact factor to account for the dynamic effects of the traffic.

Figure 2.15 is a plot of an E80 live load, prism earth load, and the sum of the two. An E80 loading is designed to represent a railway load and again this includes a 50 percent impact factor.

An H20 load consists of two 16,000-lb concentrated loads applied to two 18-in by 20-in areas, one located over the point in question and the other located at a distance of 72 in away. It is interesting to note (Fig. 2.14) that for the example considered, the minimum total load would occur at about 4½ ft of cover. Also, it is evident from Fig. 2.14 that live loads have little effect on pipe performance except at shallow depths. Thus, design precautions should be taken for shallow installations under roadways. If the live load is an impact-type load, it can be as much as twice the static surface load. However, from a practical standpoint, the impact factor will usually be less than 1.5. At extremely shallow depths of cover, a flexible pipe may deflect and rebound under dynamic loading. Special precautions should be taken for shallow burials in roadways to prevent surface breakup.

Aircraft loads

Design live loads for modern airports may be very large. Airports are often designed for wheel loads of aircraft which have not yet been designed. Table 2.7 lists live loads for an aircraft loading of 180,000-lb dual-tandem gear assembly.

In the design for live loads on pipe buried under runway pavement, the impact factor is taken as 1.0. This is because the load is partially

TABLE 2.6 Values of C_s

Values of Load Coefficients, C_s, for Concentrated and Distributed Superimposed Loads Vertically Centered Over Conduit*

$D/2H$ or $B_c/2H$	$M/2H$ or $L/2H$													
	0.1	0.2	0.3	0.4	0.5	0.6	0.7	0.8	0.9	1.0	1.2	1.5	2.0	5.0
0.1	0.019	0.037	0.053	0.067	0.079	0.089	0.097	0.103	0.108	0.112	0.117	0.121	0.124	0.128
0.2	0.037	0.072	0.103	0.131	0.155	0.174	0.189	0.202	0.211	0.219	0.229	0.238	0.244	0.248
0.3	0.053	0.103	0.149	0.190	0.224	0.252	0.274	0.292	0.306	0.318	0.333	0.345	0.355	0.360
0.4	0.067	0.131	0.190	0.241	0.284	0.320	0.349	0.373	0.391	0.405	0.425	0.440	0.454	0.460
0.5	0.079	0.155	0.224	0.284	0.336	0.379	0.414	0.441	0.463	0.481	0.505	0.525	0.540	0.548
0.6	0.089	0.174	0.252	0.320	0.379	0.428	0.467	0.499	0.524	0.544	0.572	0.596	0.613	0.624
0.7	0.097	0.189	0.274	0.349	0.414	0.467	0.511	0.546	0.584	0.597	0.628	0.650	0.674	0.688
0.8	0.103	0.202	0.292	0.373	0.441	0.499	0.546	0.584	0.615	0.639	0.674	0.703	0.725	0.740
0.9	0.108	0.211	0.306	0.391	0.463	0.524	0.574	0.615	0.647	0.673	0.711	0.742	0.766	0.784
1.0	0.112	0.219	0.318	0.405	0.481	0.544	0.597	0.639	0.673	0.701	0.740	0.774	0.800	0.816
1.2	0.117	0.229	0.333	0.425	0.505	0.572	0.628	0.674	0.711	0.740	0.783	0.820	0.849	0.868
1.5	0.121	0.238	0.345	0.440	0.525	0.596	0.650	0.703	0.742	0.774	0.820	0.861	0.894	0.916
2.0	0.124	0.244	0.355	0.454	0.540	0.613	0.674	0.725	0.766	0.800	0.849	0.894	0.930	0.956

*Influence coefficients for solution of Holl's and Newmark's integration of the Boussinesq equation for vertical stress.

SOURCE: "Design and Construction of Sanitary and Storm Sewers" Manuals and Reports on Engineering Practice No. 37, American Society of Civil Engineers and Manual of Practice No. 9, Water Pollution Control Federation, 1969, p. 206. Reprinted by permission.

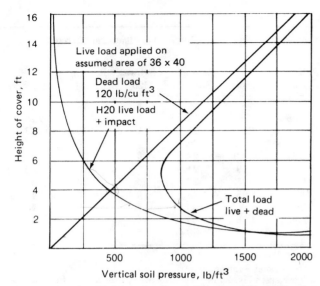

Figure 2.14 H20 highway loading. *(Reprinted from Handbook of Steel Drainage and Highway Construction Products, by permission of the American Iron and Steel Institute, Washington, D.C.)*

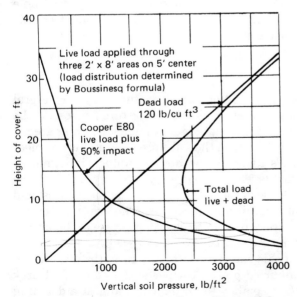

Figure 2.15 Cooper E80 Live loading. *(Reprinted from Handbook of Steel Drainage and Highway Construction Products, by permission of the American Iron and Steel Institute, Washington, D.C.)*

TABLE 2.7 Live Loads

	Live load transferred to pipe, lb/in†				Live load transferred to pipe, lb/in†		
Height of cover, ft	Highway H20*	Railway E80†	Airport‡	Height of cover, ft	Highway H20*	Railway E80†	Airport‡
1	12.50	—	—	14	§	4.17	3.06
2	5.56	26.39	13.14	16	§	3.47	2.29
3	4.17	23.61	12.28	18	§	2.78	1.91
4	2.78	18.40	11.27	20	§	2.08	1.53
5	1.74	16.67	10.09	22	§	1.91	1.14
6	1.39	15.63	8.79	24	§	1.74	1.05
7	1.22	12.15	7.85	26	§	1.39	§
8	0.69	11.11	6.93	28	§	1.04	§
10	§	7.64	6.09	30	§	0.69	§
12	§	5.56	4.76	35	§	§	§
				40	§	§	§

*Simulates 20 ton truck traffic + impact.
†Simulates 80,000 lb/ft railway load + impact.
‡180,000-lb dual-tandem gear assembly. 26 in spacing between tires and 66 in center-to-center spacing between fore and aft tires under a rigid pavement 12 in thick + impact.
§Negligible live load influence.
SOURCE: Reprinted from the *Uni-Bell Handbook*, by permission.

taken by the aircraft's wings when the aircraft is landing. For taxiways, aprons, and so on, an impact factor may be necessary (see Table 2.5). The design engineer should seek current data available from the Federal Aviation Administration.

Frost Loading

When freezing atmospheric conditions exist continuously for several hours, ice layers or lenses form as shallow soil moisture freezes. As the frost penetrates downward, additional small volumes of water freeze. This freezing has a drying effect upon the soil since the water is no longer available to satisfy the soil's attraction for capillary water. Thus, groundwater from below the frost layer is attracted by capillary action to the area of lower potential. This water also freezes as it reaches the frost and the process continues until equilibrium is reached. The freezing of ice below existing ice layers causes pressure to develop because of the expansion due to growth (volume increase) of ice.

It has been shown that this expansive pressure can substantially increase vertical loads on buried pipes. A paper authored by W. Harry Smith (*AWWA Journal*, December 1975) indicates almost a doubling of load during the deepest frost penetration. For this study, the test-pipe setup was essentially nonyielding.

The test pipe was split longitudinally in two halves and load cells

Figure 2.16 Schematic of split pipe with supporting load cell.

were placed inside the pipe (see Fig. 2.16), such that the load cell was
between the two halves. The maximum deflection of the load cells was
0.003 in. The test pipe simulated an extremely rigid pipe. Due to this
rigidity, the load increase was greatly magnified. The previously discussed spring analogy can be applied here. In this case, the test pipe is
represented by a very stiff spring and the soil side fills by softer
springs. It is clear that the stiffer spring will take most of the load.

The increase in load, due to frost penetration, is less pronounced for
flexible pipes. For example, plastic pipes such as PVC may have a
small increase in deflection without any structural distress. Normally,
designs require that pipes are placed 1 or 2 ft below the frost line. The
design engineer should be aware that frost action may increase loads
on a rigid pipe.

Loads Due to Expansive Soils

Expansive soils were mentioned briefly in the "Longitudinal Loading"
section concerning possible ground movement. Certain soils, primarily
bentonite clays, expand and contract severely as a function of moisture
content. Soil expansion can cause an increase in soil pressure just as frost
can cause an increase in soil pressure. This rise in pressure is directly
due to expansion and is a function of confinement. Tremendously high
pressures can result if such soils are confined between nonyielding surfaces. However, data are lacking concerning such forces which may be
induced on buried conduits. This lack of data can probably be attributed
to design practices that do not allow such soils to be placed directly
around the pipe. Also, in the case of gravity sewers, designs usually require such material to be removed for certain depths below the pipe if
moisture content is variable at such depths. The primary reason for this
is to ensure that grade is maintained. The design engineer should be cog-
nizant that expansive soils do pose certain potential problems. He should
seek advice from a component soils (geotechnical) engineer, and then
take appropriate steps in the installation design to mitigate adverse effects of expanding soils.

Bibliography

1. American Association of Civil Engineers and Water Pollution Control Federation, *Gravity Sanitary Sewer, Design and Construction,* 1982.
2. *Handbook of Steel Drainage and Highway Construction Products,* American Iron and Steel Institute, Chicago, 1971.
3. American Railway Engineering Assoc., Manual of Recommended Practice, AREA Spec 1-4-28, Chicago.
4. Marston, Anson, and A. O. Anderson, "The Theory of Loads on Pipes in Ditches and Tests of Cement and Clay Drain Tile and Sewer Pipe," Bulletin 31, Iowa Engineering Experiment Station, Ames, Iowa, 1913.
5. Moser, A. P., R. K. Watkins, and O. K. Shupe, "Design and Performance of PVC Pipes Subjected to External Soil Pressure," Buried Structures Laboratory, Utah State University, Logan, Utah, June 1976.
6. Newmark, N. M., "Influence Charts for Computation of Stresses in Elastic Foundations," Bulletin 338, University of Illinois Engineering Experiment Station, Urbana, Ill., 1942.
7. Piping Systems Institute, "Course Notebook," Utah State University, Logan, Utah, 1980.
8. Reissner, E., *J. of Appl. Mech.* "On Final Bending of Presserized Tubes," *Trans. ASME,* September 1959, pp. 386–392.
9. Schlick, W. J., "Loads on Pipe in Wide Ditches," Bulletin 108, Iowa Engineering Experiment Station, Ames, Iowa, April 6, 1932.
10. Spangler, M. G., "The Structural Design of Flexible Pipe Culverts," Bulletin 153, Iowa Engineering Experiment Station, Ames, Iowa, 1941.
11. Spangler, M. G., and R. L. Handy, *Soil Engineering,* Harper & Row (Intext), New York, 1973.
12. Terzaghi, K., and R. B. Beck, *Soil Mechanics in Engineering Practice,* Wiley, New York, 1967.
13. Uni-Bell PVC Pipe Association, *Handbook of PVC Pipe Design and Construction,* Dallas, Texas, 1982.
14. Watkins, R. K., and A. P. Moser, "Response of Corrugated Steel Pipe to External Soil Pressures," Highway Research Record 373, 1971, pp. 88–112.

Chapter

3

Design of Gravity
Flow Pipes

Design methods which are used to determine an installation design for buried gravity flow pipes are given in this chapter. Soil types and their uses in pipe embedment and backfill are discussed. Design methods are placed in two general classes—rigid pipe design and pressure pipe design. Pipe performance limits are given and recommended safety factors are reviewed.

The finite element method for design of buried piping systems is relatively new. The use of this powerful tool is increasing with time. A detailed discussion of this method is included.

Soils

The importance of soil density (compaction) and soil type in contributing to buried pipe performance has long been recognized by engineers. The pipe-zone backfill, which is often referred to as the soil envelope around the pipe, is most important. An introduction and a brief discussion of embedment soils is presented in Chap. 1. In this chapter, additional information on soil classification and soil-pipe interaction is provided.

Soil classes

Professor Arthur Casagrande proposed a soil classification system for roads and airfields in the early 1940s. This system, now called the Unified Soil Classification System (USCS), has been adopted by many groups and agencies including the Army Corps of Engineers and the Bureau of Reclamation. ASTM's version of the United Soil Classification System is entitled "Classification of Soils for Engineering Purposes" and carries the designation D2487.

The USCS is based on the textural characteristics for those soils with a small amount of fines such that the fines have little or no influence on behavior. For those soils where fines affect the behavior, classification is based on plasticity-compressibility characteristics. The plasticity-compressibility characteristics are evaluated by plotting the plasticity chart. The position of the plotted points yields classification information. The following properties form the basis of soil classification and identification:

1. Percentages of gravel, sand, and fines [fraction passing 0.75 mm (no. 200) sieve]

2. Shape of grain-size distribution curve (see Fig. 3.1)

3. Plasticity and compressibility characteristics (see Fig. 3.2)

A soil is given a descriptive name and letter symbol to indicate its principal characteristics. (See ASTM D 2487 or any text on soil mechanics.)

Embedment materials listed here include the soil types defined according to the USCS and a number of processed materials.

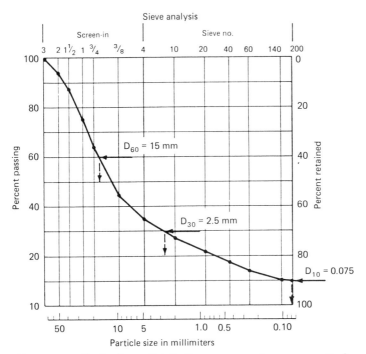

Figure 3.1 Grain-size distribution curve for a particular soil. *(Reprinted, by permission, from ASTM Standards D2487, Fig. 4.)*

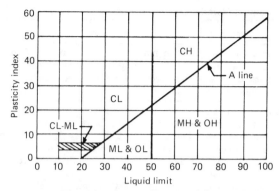

Figure 3.2   Plasticity chart.   *(Reprinted, by permission, from Asphalt Institute Soils Manual.)*

ASTM D 2321 "Underground Installation of Flexible Thermoplastic Sewer Pipe" breaks down embedment materials into five classes. These classes along with the USCS letter designation and description are given in Table 3.1.

Class I.   Angular, ¼ to 1½ in (6 to 40 mm) graded stone, including a number of fill materials that have regional significance such as coral, slag, cinders, crushed shells, and crushed stone

Note: The size range and resulting high voids ratio of class I material make it suitable for use to dewater trenches during pipe installation. This permeable characteristic dictates that its use be limited to locations where pipe support will not be lost by migration of fine grained natural material from the trench walls and bottom or migration of other embedment materials into the class I material. When such migration is possible, the material's minimum size range should be reduced to finer than ¼ in (6 mm) and the gradation properly designed to limit the size of the voids

Class II.   Coarse sands and gravels with maximum particle size of 1½ in (40 mm), including variously graded sands and gravels containing small percentages of fines, generally granular and noncohesive, either wet or dry. Soil Types GW, GP, SW, and SP are included in this class.

Sands and gravels that are clean or borderline between clean and with fines should be included. Coarse-grained soils with less than 12 percent but more than 5 percent fines are neglected in ASTM D2487 and the USCS and should be included. The gradation of class II material influences its density and pipe-support strength when loosely placed. The gradation of class II material may be critical to the pipe

TABLE 3.1   Description of Embedment Material Classifications

Soil class	Soil type	Description of material classification
Class I soils*	—	Manufactured angular, granular material, ¼ to 1½ inches (6 to 40 mm) size, including materials having regional significance such as crushed stone or rock, broken coral, crushed slag, cinders, or crushed shells.
Class II soils†	GW	Well-graded gravels and gravel-sand mixtures, little or no fines. 50% or more retained on no. 4 sieve. More than 95% retained on no. 200 sieve. Clean.
	GP	Poorly graded gravels and gravel-sand mixtures, little or no fines. 50% or more retained on no. 4 sieve. More than 95% retained on no. 200 sieve. Clean.
	SW	Well-graded sands and gravelly sands, little or no fines. More than 50% passes no. 4 sieve. More than 95% retained on no. 200 sieve. Clean.
	SP	Poorly graded sands and gravelly sands, little or no fines. More than 50% passes no. 4 sieve. More than 95% retained on no. 200 sieve. Clean.
Class III soils‡	GM	Silty gravels, gravel-sand-silt mixtures. 50% or more retained on no. 4 sieve. More than 50% retained on no. 200 sieve.
	GC	Clayey gravels, gravel-sand-clay mixtures. 50% or more retained on no. 4 sieve. More than 50% retained on no. 200 sieve.
	SM	Silty sands, sand-silt mixtures. More than 50% passes no. 4 sieve. More than 50% retained on no. 200 sieve.
	SC	Clayey sands, sand-clay mixtures. More than 50% passes no. 4 sieve. More than 50% retained on no. 200 sieve.
Class IV soils	ML	Inorganic silts, very fine sands, rock flour, silty or clayey fine sands. Liquid limit 50% or less. 50% or more passes no. 200 sieve.
	CL	Inorganic clays of low to medium plasticity, gravelly clays, sandy clays, silty clays, lean clays. Liquid limit 50% or less. 50% or more passes no. 200 sieve.
	MH	Inorganic silts, micaceous or diatomaceous fine sands or silts, elastic silts. Liquid limit greater than 50%. 50% or more passes no. 200 sieve.
	CH	Inorganic clays of high plasticity, fat clays. Liquid limit greater than 50%. 50% or more passes no. 200 sieve.
Class V soils	OL	Organic silts and organic silty clays of low plasticity. Liquid limit 50% or less. 50% or more passes no. 200 sieve.
	OH	Organic clays of medium to high plasticity. Liquid limit greater than 50%. 50% or more passes no. 200 sieve.
	PT	Peat, muck and other highly organic soils.

*Soils defined as Class I materials are not defined in ASTM D2487.
†In accordance with ASTM D2487, less than 5% pass no. 200 sieve.
‡In accordance with ASTM D2487, more than 12% pass no. 200 sieve. Soils with 5% to 12% pass no. 200 sieve fall in borderline classification, e.g., GP-GC.
SOURCE: Reprinted by permission. Copyright ASTM.

support and stability of the foundation and embedment if the material is imported and is not native to the trench excavation. A gradation other than well-graded, such as uniformly graded or gap graded, may permit loss of support by migration into void spaces of a finer grained natural material from the trench wall and bottom.

Class III. Fine sand, and clayey (clay filled) gravels, including fine sands, sand-clay mixtures and gravel-clay mixtures. Soil types GM, GC, SM, and SC are included in this class.

Class IV. Silt, silty clays, and clays, including inorganic clays and silts of low to high plasticity and liquid limits. Soil types MH, ML, CH, and CL are included in this class.

Note: Caution should be used in the design and selection of the degree and method for compaction for class IV soils because of the difficulty in properly controlling the moisture content under field conditions. Some class IV soils with medium to high plasticity and with liquid limits greater than 50 percent (CH, MH, CH-MH) exhibit reduced strength when wet and should only be used for bedding, haunching, and initial backfill in arid locations where the pipe embedment will not be saturated by groundwater, rainfall, and/or exfiltration from the pipeline system. Class IV soils with low to medium plasticity and with liquid limits lower than 50 percent (CL, ML, CL-ML) also require careful consideration in design and installation to control moisture content, but need not be restricted in use to arid locations.

Class V. This class includes the organic soils OL, OH, and PT as well as soils containing frozen earth, debris, rocks larger than 1½ inch (40 mm) in diameter, and other foreign materials. These materials are not recommended for bedding, haunching, or initial backfill.

Soil-pipe interaction

Design of a buried conduit has a basic objective of adequate overall performance at minimum cost. Overall performance includes not only structural performance, but also service life. Minimum cost should include all costs including lifetime maintenance.

Initial cost is often broken down into piping material cost and installation cost. Various pipe products have differing strengths and stiffness characteristics and may require differing embedment materials and placement techniques depending on the in situ soil and depth of burial. Products which allow for minimum initial or installation costs may not be the lowest cost alternative when consideration is given to total lifetime cost. Soil-structure interaction influences pipe performance and is a function of both the pipe properties and embedment soil properties, and therefore impacts total system costs. The design engineer should consider soil-structure interaction in the installation design and lifetime cost estimates.

The soil-pipe system is highly statically indeterminate. This means that the interface pressure between the soil and pipe cannot be calculated by statics alone—the stiffness properties of both soil and pipe must also be considered. The ratio of pipe stiffness to soil stiffness (PS/E') determines to a large degree the load imposed on the conduit. For example, a so-called "rigid pipe" will have a much greater load than a "flexible pipe" installed under the same or similar conditions.

Soil to be placed in the pipe zone should be capable of maintaining the specified soil density. Also, to eliminate pressure concentrations, the soil should be uniformly placed and compacted around the pipe.

Various placement methods can be used depending upon system parameters such as soil type, required density, burial depth, pipe stiffness, and pipe strength. The following are suggested as methods which will achieve desirable densities with the least effort (see Table 3.2).

Certain manufactured materials may be placed by loose dumping with a minimum of compactive effort. These materials must be angular and granular such as broken coral, crushed stone or rock, crushed shells, crushed slag, or cinders and have a maximum size of 1½ in (40 mm). Care should be taken to assure proper placement of these materials under pipe haunches.

With coarse-grained soils containing less than 5 percent fines such as GW, GP, SW, SP, GW-GP, and SW-SP, the maximum density will be obtained by compacting by saturation or vibration. If internal vibrators are used, the height of successive lifts or backfill should be limited to the penetrating depth of the vibrator. If surface vibrators are used, the backfill should be placed in lifts of 6 to 12 in (150 to 300 mm). This material may also be compacted by hand tamping or other means provided that the desired relative density is obtained.

Coarse-grained soils which are borderline between clean and those with fines containing between 5 percent and 12 percent fines, such as GW-GM, SW-SM, GW-GC, SW-SC, GP-GM, SP-SM, GP-GC, and SP-SC, should be compacted either by hand or mechanical tamping, saturation or vibration, or whichever method meets the required density.

Coarse-grained soils containing more than 12 percent fines, such as GM, GC, SM, SC, and any borderline cases in the group (that is, GM-SM), should be compacted by hand or mechanical tamping. The backfill should be placed in lifts of 4 to 6 in (100 to 150 mm). Fine-grained soils such as MH, CH, ML, CL, SC-CL, SM-ML, and ML-CL, should be compacted by hand or mechanical tamping in lifts of 4 to 6 in (100 to 150 mm).

Rigid Pipe Analysis

Three-edge bearing strength

Rigid nonpressure pipes are of four general types and are covered by four ASTM specifications. Asbestos-cement pipe is governed by ASTM C428. Vitrified clay pipe is specified in two-strength class by ASTM C700. Nonreinforced concrete pipe is covered by ASTM C14. Reinforced concrete nonpressure pipe is specified by its so-called D-load strength as given in ASTM C76. The D-load is the three-edge bearing strength divided by diameter.

Designs of the various types of rigid pipe for nonpressure applications to resist external loads require knowledge of available pipe strengths as well as the construction or installation conditions to be encountered. Rigid pipe is tested for strength in the laboratory by the three-edge bearing test (see Fig. 3.3 for diagram of test). Methods for testing are described in detail in the respective ASTM specifications for the specific pipe product. The three-edge bearing strength is the load per length (usually pounds per foot) required to cause crushing or critical cracking of the pipe test specimen. This strength is the load at failure in a testing machine. It is not necessarily the load that will cause failure in the soil.

Bedding factors and classifications

In Chap. 2, we learned that for rigid pipe the soil load can be calculated by Marston's equation ($W_d = C_d \gamma B_d^2$). Experience has shown that the Marston load to cause failure is usually greater than the three-edge bearing strength and depends on how the pipe was bedded (Fig. 3.3).

The Marston load to cause failure is called the field strength. The ratio of field strength to three-edge bearing strength is termed the "bedding factor" since it is dependent upon how the pipe was bedded (installed). The term "bedding factor" as used by Marston is sometimes called the "load factor." The two terms have reference to the same parameter and may be used interchangeably.

$$\text{Bedding factor} = \frac{\text{field strength}}{\text{three-edge bearing stength}} \qquad (3.1)$$

Major pipe manufacturing associations recommend bedding factors which correspond to those listed in the Water Pollution Control Fed-

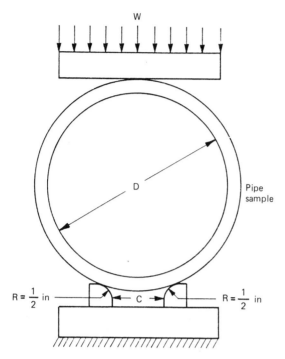

Figure 3.3   Three-edge bearing (wood block) schematic.

eration Manual of Practice no. FD-5, "Gravity Sanitary Sewer Design and Construction."

These bedding types (classes) are shown in Fig. 3.4 and corresponding bedding factors (load factors) are given in Table 3.2.

Installation design

Equation (3.1) may be solved for the three-edge bearing strength as follows:

$$\text{Three-edge bearing strength} = \frac{\text{field strength}}{\text{bedding factor}} \qquad (3.2)$$

TABLE 3.2   Bedding Factors

Bedding class	Load factor
A	2.8–3.4
B	1.9
C	1.5
D	1.1

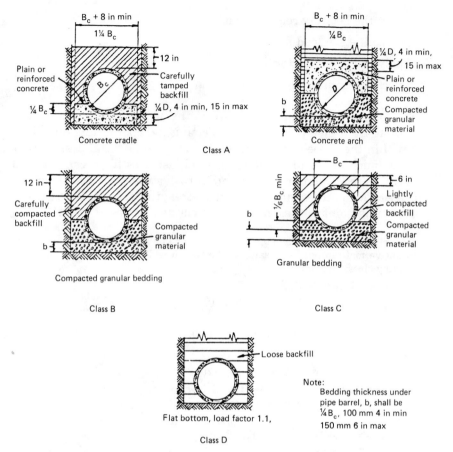

Figure 3.4 Class of bedding for rigid sewer pipes. [*Note*: In rock trenches, excavate at least 6-in below bell. In unstable material, such as peat or expansive soil, remove unstable material and replace with a select fill material (consult competent soils engineer).] *(Reprinted with permission of Water Pollution Control Federation.)*

The field strength is the Marston load that will cause failure in the field. Most designers and specifications require a factor of safety in the design. Thus the required strength is as follows:

$$\frac{\text{Required three-edge}}{\text{bearing strength}} = \frac{\text{design load} \times \text{factor of safety}}{\text{bedding factor}} \qquad (3.3)$$

A design procedure to select the appropriate pipe classification or strength is outlined as follows:

1. Determine the earth load
2. Determine the live load

3. Select the bedding requirement
4. Determine the load factor
5. Apply the safety factor
6. Select the appropriate pipe strength

The following example will illustrate the use of the six design steps and basic rigid pipe principles in selecting the appropriate pipe. It is not within the scope of this text to discuss pipe material design [that is, how much material (reinforcing steel, asbestos fiber, cement, and so forth) is required to meet a specific crush strength is not included]. Since ASTM specifies minimum crush strengths, these will be used as a beginning point for design in the discussion here.

Example 3.1   A 15 in diameter sanitary sewer line is to be installed 14 ft deep in native material which is sand. If the trench width is 3.0 ft, what pipe and bedding classes should be selected?

1. *Determine earth load*

$$\frac{H}{B_d} = \frac{14}{3} = 4.67$$

sand, $K\mu$ = 0.165

From Fig. 2.2, C_d = 2.4

$$W_d = C_d \gamma B_d^2 = 2.4\,(120)(3.0)^2 = 2592 \text{ lb/ft}$$

2. *Determine live load* (assume H20 highway loading). From Fig. 2.14, it is noted that the live load is negligible for 14 ft of cover.

3. *Select bedding.* Economic and practical engineering judgment is required. Compare class D, C, and B (Fig. 3.4).

4. *Determine load factor*

Class D   Table 3.2:   BF = 1.1
Class C   BF = 1.5
Class B   BF = 1.9

5. *Apply safety factor*

Concrete (ACPA)   Recommendations: SF = 1.25–1.5
Reinforced concrete (ACPA)   SF = 1.0 based on 0.01" crack

Clay (CPI)   SF = 1.0–1.5
Asbestos cement (ACPPA)   SF = 1.0–1.5

6. *Select pipe strength*

$$W_{3\text{-edge}} = W_c \times \frac{\text{SF}}{\text{BF}}$$

$$W_{D\text{-load}} = W_c \times \frac{SF}{BF} \times D$$

$$W_{3\text{-edge}} = 2592 \frac{SF}{BF}$$

$$W_{D\text{-load}} = 2074 \frac{SF}{BF}$$

Minimum Required Strength for SF = 1.5

Bedding class	Three-edge (lb/ft)	D-load (lb/ft)/ft
B	2046	1637
C	2592	2074
D	3535	2828

Choice may be based on job details including economic consideration of pipe versus bedding cost. Choose a strength class that equals or exceeds strengths given in table above.

Example 3.2 Suppose the trench width of 3.0 ft cannot be maintained at the top of the pipe in Example 3.1. What are the required strengths if the transition trench width is reached?

1. Determine transition width

$$\frac{H}{B_c} = \frac{14}{1.25} = 11.20$$

$$r_{sdp} = 0.5$$

From Fig. 2.6, $B_d/B_c = 2.9$ and

$$B_d(\text{transition}) = \frac{B_d}{B_c} \cdot B_c = 2.9 \left(\frac{15}{12}\right) = 3.6 \text{ ft}$$

2. Determine earth load

$$\frac{H}{B_d} = \frac{14}{3.6} = 3.9$$

granular, $K\mu = 0.192$

From Fig. 2.2, $C_d = 2.0$ and

$$W_d = C_d\gamma B_d^2 = 2.0 \,(120)(3.6)^2 = 3110 \text{ lb/ft}$$

Alternately,

$$\frac{H}{B_c} = \frac{14}{1.25} = 11.20$$

$$r_{sdp} = 0.5$$

From Table 2.2,

$$C_c = 1.5 \frac{H}{B_c} - 0.07 = 16.73 \approx 16.7$$

$$W_c = C_c\gamma B_c^2 = 16.7 \,(120)(1125)^2 = 3131 \text{ lb/ft}$$

At the transition width, $W_c \approx W_d$

$$W_{3\text{-edge}} = 3131 \frac{\text{SF}}{\text{LF}}$$

$$W_{D\text{-load}} = 2505 \frac{\text{SF}}{\text{LF}}$$

Minimum Required Strength for SF = 1.5

Bedding class	Three-edge (lb/ft)	D-load [lb/(ft)/(ft)]
B	2472	1977
C	3131	2505
D	4270	3416

Flexible Pipe Analysis

Installation design

Three parameters are most essential in the design or the analysis of any flexible conduit installation. They are

1. Load (depth of burial)
2. Soil stiffness in pipe zone
3. Pipe stiffness

The design load on the pipe is easily calculated using the prism-load theory as discussed in Chap. 2. This load is simply the product of the soil unit weight and the height of cover. Research has shown that the long-term load on a flexible pipe approaches the prism load. Thus, if this load is used in design, the deflection lag factor should be taken as unity.

The soil stiffness is usually expressed in terms of E' (effective soil modulus, lb/in^2). E' is dimensionally load per square inch and is a soil modulus term. The soil modulus (E') is a function of soil properties such as soil density, soil type, and moisture content. Experience has shown that soil density is the most important parameter influencing soil stiffness.

For flexible pipes, pipe stiffness rather than crush strength is usually the controlling pipe material property. Pipe stiffness may be expressed in terms of various parameters as follows:

Pipe Stiffness Terminology

Stiffness factor = EI

Ring stiffness = EI/r^3 (or sometimes EI/D^3)

Pipe stiffness = $F/\Delta y = 6.7 \; EI/r^3$

where E = modulus of elasticity, lb/in^2
I = moment of inertia of the wall cross-section per unit length of pipe, $\text{in}^4/\text{in} = \text{in}^3$
r = mean radius of pipe, in
D = mean diameter of pipe, in
F = force, lb/in
Δy = vertical deflection, in

The most commonly used terminology is pipe stiffness $(F/\Delta y)$. For a given pipe product, this term is readily determined in the laboratory by a parallel plate loading test. In this test, a pipe sample is placed between two horizontal parallel plates in a testing machine. A compressive load is applied and increased until the vertical deflection (Δy) reaches 5 percent of the diameter. $F/\Delta y$ is the load at 5 percent divided by the sample length and divided by the vertical deflection Δy. Typical units for $F/\Delta y$ are lb/in^2. This is evident from the third equation in the above table as it is clear that $F/\Delta y$ has the same units as the modulus of elasticity (E).

In summary, the three most important parameters for flexible pipe analysis and design are (1) load, (2) soil stiffness, and (3) pipe stiffness. Any design method that does not include a consideration of these three parameters is incomplete.

For a flexible pipe, vertical deflection is the variable that must be controlled by proper installation design. This deflection is a function of the three parameters discussed above.

Spangler's Iowa formula

M. G. Spangler, a student of Anson Marston, observed that the Marston theory for calculating loads on buried pipe was not adequate for flexible pipe design. Spangler noted that flexible pipes provide little inherent stiffness in comparison to rigid pipes, yet they perform remarkably well when buried in soil. This significant ability of a flexible pipe to support vertical soil loads is derived from (1) the redistribution of loads around the pipe, and (2) the passive pressures induced as the sides of the pipe move outward against the surrounding earth. These considerations, coupled with the idea that the ring deflection may form a basis for flexible pipe design, prompted M. G. Spangler to study flexible pipe behavior to determine an adequate design procedure. His research and testing led to the derivation of the Iowa formula which he published in 1941.[22]

Spangler incorporated the effects of the surrounding soil on the pipe's deflection. This was accomplished by assuming that Marston's theory of loads applied, and that this load would be uniformly distributed at the plane at the top of the pipe. He also assumed a uniform pressure over part of the bottom, depending upon the bedding angle. On the sides, he assumed the horizontal pressure h on each side would be proportional to the deflection of the pipe into the soil. The constant of proportionality was defined as shown in Fig. 3.5 and was called the modulus of passive resistance of the soil. The modulus would presumably be a constant for a given soil and could be measured in a simple lab test. Through analysis he derived the Iowa formula as follows:

$$\Delta X = \frac{D_L K W_c r^3}{EI + 0.061er^4} \tag{3.4}$$

where D_L = deflection lag factor
K = bedding constant
W_c = Marston's load per unit length of pipe, lb/in
r = mean radius of the pipe, in
E = modulus of elasticity of the pipe material, lb/in^2
I = moment of inertia of the pipe wall per unit length, in^4/in = in^3
e = modulus of passive resistance of the side fill, lb/(in^2)(in)
ΔX = horizontal deflection or change in diameter, in

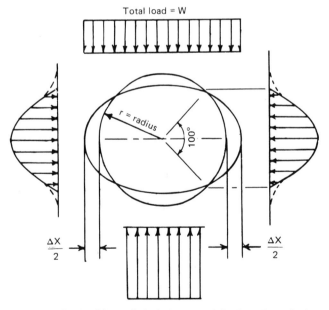

Figure 3.5   Basis of Spangler's derivation of the Iowa formula for deflection of buried pipes. [$\Delta X = D_L K W_c r^3/(EI + 0.061er^4)$ (the Iowa formula); where $e = 2h/\Delta X$; $2r = D$ = pipe diameter; K = Bedding constant; D_L = Deflection lag factor; and EI = Stiffness factor (related to pipe stiffness).] *(Reprinted with permission of Utah State University.)*

Equation (3.4) can be used to predict deflections of buried pipe if the three empirical constants K, D_L, and e are known. The bedding constant, K, accommodates the response of the buried flexible pipe to the opposite and equal reaction to the load force derived from the bedding under the pipe. The bedding constant varies with the width and angle of the bedding achieved in the installation. The bedding angle is shown in Fig. 3.6. Table 3.3 contains a list of bedding factors, K, dependent upon the bedding angle. These were determined theoretically by Spangler and published in 1941. As a general rule, a value of $K = 0.1$ is assumed.

In 1958, Reynold K. Watkins, a graduate student of Spangler, was investigating the modulus of passive resistance through model studies and examined the Iowa formula dimensionally.[31] The analysis determined that e could not possibly be a true property of the soil in that its dimensions are not those of a true modulus. As a result of Watkins' effort, another soil parameter was defined. This was the modulus of soil reaction, $E' = er$. Consequently, a new formula called the modified Iowa formula was proposed.

$$\Delta X = \frac{D_L \, KW_c r^3}{EI + 0.061E'r^3} \tag{3.5}$$

Two other observations from Watkins' work are of particular note. (1) There is little point in evaluating E' by a model test and then using this modulus to predict ring deflection as the model gives ring deflection directly. (2) Ring deflection may not be the only performance limit.

Another parameter in the Iowa formula, needed to calculate deflections, is the deflection lag factor D_L. Spangler recognized that in soil-pipe systems, as with all engineering systems involving soil, the soil consolidation at the sides of the pipe continues with time after installation of the pipe. His experience had shown that deflections could increase by as much as 30 percent over a period of 40 years. For this reason, he recommended the incorporation of a deflection lag factor of

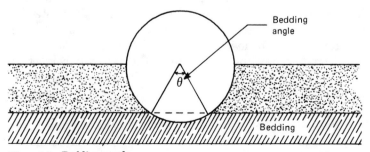

Figure 3.6   Bedding angle.

TABLE 3.3   Values of Bedding Constant, K

Bedding angle, degrees	K
0	0.110
30	0.108
45	0.105
60	0.102
90	0.096
120	0.090
180	0.083

1.5 as a conservative design procedure. However, recall that the load proposed by Spangler was the Marston load for a flexible pipe. For most sewer pipe installations the prism load is at least 1.5 times greater than the Marston load (see Chap. 2 for soil loads on pipe). If the prism load is used for design, a design deflection lag factor $D_L = 1.0$ should be used.

The remaining parameter in the modified Iowa formula is the soil modulus E'. Many research efforts have attempted to measure E' without success. The most useful method has involved the measurement of deflections of a buried pipe for which installation conditions are known, followed by a back calculation through the Iowa formula to determine the effective value of E'. This requires assumed values for the load, the bedding factor, and the deflection lag factor. Inconsistent assumptions have led to a variation in reported values of E'.

One attempt to acquire information on values of E' was conducted by Amster K. Howard of the United States Bureau of Reclamation.[5] Howard used data from laboratory and field tests to compile a table of average E' values for various soil types and densities (see Table 3.4). He assigned values to E', K, and W_c and then used the Iowa formula to calculate a theoretical value of deflection. This theoretical deflection was then compared with actual measurements. By assuming the E' values of Table 3.4 and a bedding constant $K = 0.1$, Howard was able to correlate the theoretical and empirical results to within ± 2 percent deflection when he used the prism soil load. This means that if theoretical deflections using Table 3.4 were approximately 5 percent, measured deflections would range between 3 and 7 percent. Mr. Howard is reported to have used a deflection lag factor $D_L = 1.5$ in his calculations. However if the prism load was used as reported, a lag factor $D_L = 1.0$ would have to have been used to be theoretically correct. In any case, the data in Table 3.4 is consistent with field and laboratory data taken over a 20-year period at Utah State University if the prism load is used along with a value of 1.0 for the deflection lag factor. Although the vast majority of data from Howard's study were taken from tests on steel and reinforced plastic mortar pipe with diameters greater than 24 in, it does provide some useful information to guide

TABLE 3.4   Average Values of Modulus of Soil Reaction, E' (For Initial Flexible Pipe Deflection)

Soil type-pipe bedding material (Unified Classification System*)	Dumped	E' for degree of compaction of bedding, lb/in^2		
		Slight, < 85% proctor, < 40% relative density	Moderate, 85%–95% proctor, 40%–70% relative density	High, >95% proctor, > 70% relative density
Fine-grained soils (LL > 50)† Soils with medium to high plasticity CH, MH, CH-MH	No data available; consult a competent soils engineer; Otherwise use E' = 0			
Fine-grained soils (LL < 50) Soils with medium to no plasticity CL, ML, ML-CL, with less than 25% coarse-grained particles	50	200	400	1000
Fine-grained soils (LL < 50) Soils with medium to no plasticity CL, ML, ML-CL, with more than 25% coarse-grained particles Coarse-grained soils with fines GM, GC, SM, SC contains more than 12% fines	100	400	1000	2000
Coarse-grained soils with little or no fines GW, GP, SW, SP‡ contains less than 12% fines	200	1000	2000	3000
Crushed rock	1000	3000	3000	3000
Accuracy in terms of percentage deflection§	± 2	± 2	± 1	± 0.5

*ASTM Designation D2487, USBR Designation E-3
†LL = liquid Limit
‡Or any borderline soil beginning with one of these symbols (i.e., GM-GC, GC-SC)
§For ± 1% accuracy and predicted deflection of 3%, actual deflection would be between 2% and 4%.
NOTE: Values applicable only for fills less than 50 ft (15 m). Table does not include any safety factor. For use in predicting initial deflections only, appropriate deflection lag factor must be applied for long-term deflections. If bedding falls on the borderline between two compaction categories, select lower E' value or average the two values. Percentage proctor based on laboratory maximum dry density from test standards using about 12,500 ft-lb/ft^3 (598,000 J/m^1) (ASTM D698, AASHO T-99, USBR Designation E-11). 1 lb/in^2 = 6.9 kN/m^2.
SOURCE: Amster K. Howard, "Soil Reaction for Buried Flexible Pipe," U.S. Bureau of Reclamation, Denver, Colo. Reprinted with Permission from American Society of Civil Engineers *J. Geotech. Eng. Div.*, January 1977, pp. 33–43.

designers of all flexible pipe, including PVC pipe, since it helps to give an understanding of the Iowa deflection formula.

Deflection lag and creep

The length of time that a buried flexible pipe will continue to deflect after the maximum imposed load is realized is limited. This time is a function of soil density in the pipe zone. The higher the soil density at the sides of the pipe, the shorter the time during which the pipe will continue to deflect, and the total deflection in response to the load will be less. Conversely, for lower soil densities, the creep time is longer and the resulting deflection due to creep is larger.

After the trench load reaches a maximum, the soil-pipe system continues to deflect only as long as the soil around the pipe is in the process of densification. Once the embedment soil has reached the density required to support the load, the pipe will not continue to deflect.

The full load on any buried pipe is not reached immediately after installation unless the final backfill is compacted to a high density. The increase in load with time is the largest contribution to time-dependent deflection. However, for a flexible pipe, the long-term load will not exceed the prism load. Therefore, for design, the prism load should be used, which effectively compensates for the time-dependent increase in load with trench consolidation and the resulting time-dependent deflection. Thus, when deflection calculations are based on the prism load, the deflection lag factor, D_L, should be 1.0.

Creep is normally associated with the pipe material and is defined as continuing deformation with time when the material is subjected to a constant load. Most plastics exhibit creep. As temperature increases, the creep rate under a given load increases. Also, as stress increases, the creep rate for a given temperature increases. Materials that creep are also subject to stress relaxation. Stress relaxation is defined as the decrease in stress, with time, in a material held in constant deformation. Figure 3.7 shows stress relaxation curves for PVC pipe samples held in a constant deflection condition. It is evident that stresses in PVC pipes do relax with time.

Figure 3.8 shows long-term data for buried PVC pipe. Long-term deflection tests were run at Utah State University by imposing a given soil load that was held constant throughout the duration of the test. PVC pipe material creep properties have little influence on deflection lag, but soil properties such as density exhibit great influence.

Temperature controlled tests of buried PVC pipe were run to determine the temperature effect on the long-term behavior. Data from these tests are given in graphical form in Fig. 3.9. The following procedures were used in conducting these tests. The pipe to be tested was

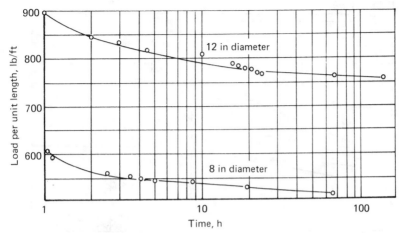

Figure 3.7 Stress relaxation curves.

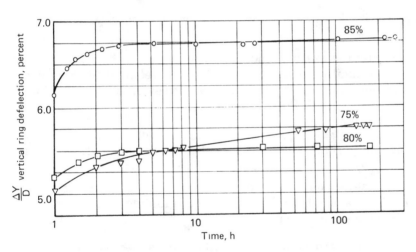

Figure 3.8 PVC pipe creep response.

placed in the load cell. It was then embedded in soil which was compacted to the specified percentage of proctor density. The load on the soil was then increased until the desired starting vertical deflection of the pipe was reached. At this point, the load as well as the temperature was held constant, and the resulting time-dependent deflection was determined. The starting deflections are somewhat arbitrary. Four of these tests were begun at about 4.75 percent deflection and two were begun between 9 and 9.5 percent deflection. The loads required to produce these deflections were different in each case.

It should be noted that for the temperature range tested, an equi-

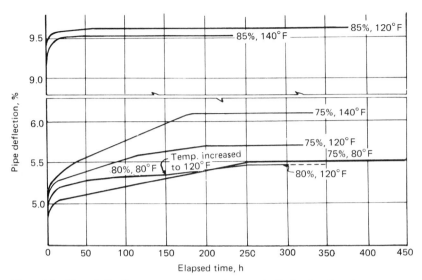

Figure 3.9 Time deflection curves—temperature controlled soil cell test.

librium state is reached and the pipe does not deflect beyond that point. The limiting deflection and the time required to reach it are largely controlled by the soil density. However, it is interesting to note the following in Fig. 3.9 for tests at different temperatures with the same soil density:

1. The equilibrium deflection is slightly larger for higher temperatures because the effective pipe stiffness is lower.

2. The time for equilibrium to be reached is shorter for higher temperatures since the soil-pipe system can interact at a faster rate in achieving equilibrium.

The above described long-term tests were carried out in a soil cell. The imposed load on a pipe in a soil cell is almost instantaneous due to the fact that the loading plane is only about 30 in above the pipe. This provides a significant advantage over tests in either trench or embankment conditions. In both the trench and the embankment, it takes substantial time for the full load to reach the pipe—months and years have been reported. When long-term tests are carried out in trenches and embankments, the change in deflections with time is due to increasing loads and soil consolidation. Figure 3.10 shows long-term deflection curves for PVC pipe buried in an embankment. The change in deflection, with respect to time, for this embankment condition is greater than that measured in soil cell tests. This time-dependent deflection is due to the increasing load that is taking place

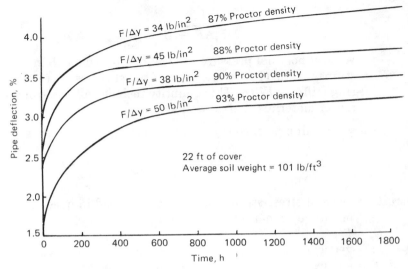

Figure 3.10   Time deflection curves—embankment test.

in the embankment tests. Whereas, in the soil cell tests, the load is applied to soil just over the pipe and is held constant. The equilibrium deflections, being approached by the curves in Fig. 3.10, are the same deflections which would result if similar pipes were tested in the soil cell at the same soil pressure and with the same soil density.

Extensive research has established that any buried flexible pipe (that is, steel, fiberglass, plastic) will continue to deflect as long as the surrounding soil consolidates. Thus, as previously stated, the creep properties of pipe materials have little effect on the long-term deflection behavior of flexible pipe when buried in soil and, in most cases, a deflection lag factor, D_L, of 1.5 conservatively accounts for long-term effects of soil consolidation. Alternatively, design can be based upon the anticipated prism load and a D_L of 1.0.

Watkins' soil-strain theory

A number of variations of Spangler and Watkins' modified Iowa formula have been proposed. All of them can be represented in the simple terms as follows:

$$\text{Deflection} = \frac{\text{load}}{\text{pipe stiffness} + (\text{constant}) (\text{soil stiffness})} \quad (3.6)$$

Upon analyzing data from many tests, Watkins wrote the Iowa formula in terms of dimensionless ratios as follows:

$$\frac{\Delta Y}{D} = \frac{PR_s}{E_s AR_s + B} \tag{3.7}$$

vertical nominal pressure at level of top of pipe, lb/in^2
R_s = stiffness ratio (This is the ratio of soil stiffness, E_s to pipe-
ring stiffness EI/D^3. This quantity includes all the properties of materials, soil as well as pipe.)

Since for a solid wall pipe of constant cross section, $I = t^3/12$, then

$$R_s = 12\frac{E_s(D)^3}{E(t)} \tag{3.8}$$

where E_s = slope of stress-strain curve for soil at load in question in
one-dimensional consolidation test
$E_s = P/\epsilon$
ϵ = vertical soil strain
A, B = empirical constants which include such terms as D_L and K
of Iowa formula

Through transposition, Eq. (3.7) can be restated

$$\frac{\Delta Y}{D\epsilon} = \frac{R_s}{AR_s + B} \tag{3.9}$$

In this form, the above equation represents a simple relationship between two-dimensionless variables: ring deflection ratio $\Delta y/(D\epsilon)$ and
stiffness ratio R_s. Figure 3.11 represents the design curve that can be

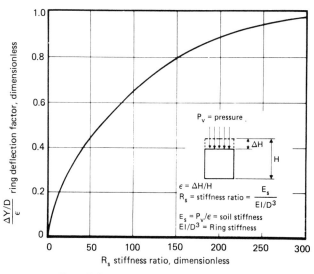

Figure 3.11   Ring deflection factor as a function of stiffness ratio.

used for predicting ring deflection. It is based on current theoretical as well as empirical data generated in Europe and America.

In most flexible pipe installations, the pipes are relatively flexible compared to recommended sidefill. Thus, the pipe follows the soil down and the deflection ratio approaches unity. The stiffness ratio, R_s, is usually greater than 300 which is to the right of the plot of Fig. 3.11. Even if R_s is usually greater than 300, it is conservative to assume $(\Delta Y/D)/\epsilon$. So the ring deflection becomes

$$\frac{\Delta Y}{D} = \epsilon \qquad (3.10)$$

This demonstrates that flexible pipe is deflected down about as much as the sidefill settles. The vertical soil strain in the fill depends upon the soil compressibility and the nominal load (Fig. 3.12). Curves shown in Fig. 3.13 relate soil strain to the soil pressure.

The use of soil strain to predict pipe deflection then becomes a simple exercise. The ratio of pipe deflection to soil strain can be determined from Fig. 3.11. This value will usually be unity for most flexible pipe installations. The load on the pipe is calculated using the prism (embankment) load theory; and the soil strain can be determined from Fig. 3.13.

For the soil to be used as embedment, a series of simple laboratory tests can be run to produce data similar to that which are shown in Fig. 3.13. However, experience has shown that data given in Fig. 3.13 are representative of most soils and can be used for design. Thus it is evident that soil density is the most important parameter in limiting pipe deflection.

Empirical method

Each of the methods discussed so far for determining load and deflection has a theoretical basis, and except for the prism load theory, all require experimental investigation to determine the unknown con-

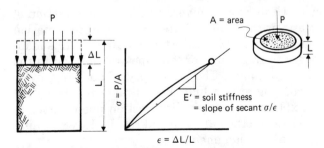

Figure 3.12 Concept for predicting settlement of soil by means of stress strain compression data from field or laboratory.

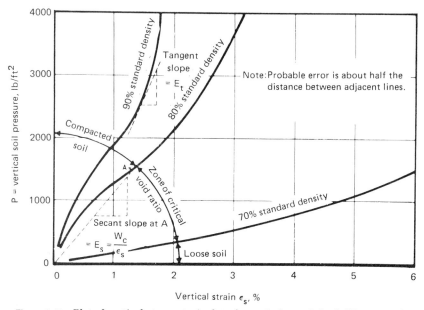

Figure 3.13   Plot of vertical stress strain data for typical trench backfill (except clay) from actual tests.

stants. In the past several years, techniques have evolved whereby a model or prototype pipe is tested until failure occurs and the total performance of the pipe is studied. Suppose a pipe is to be designed with a certain earth cover in an embankment. Without a pipe in place, no arching occurs and the soil pressure at any height is easily calculated (the prism load theory at that depth). When a pipe having good flexibility is in place, the static pressure will not be greater than the prism load pressure applied. Trying to calculate the actual pressure has frustrated researchers for years. If a pipe is installed in a prism loaded condition (for example, soil cell), the resulting deformation can be monitored without the need to calculate actual static pressure.

This procedure has been used with great success at various research laboratories such as at Utah State University under the direction of Reynold K. Watkins and at the United States Bureau of Reclamation under the direction of Amster K. Howard. Data obtained in this manner can be used directly in the design of soil-pipe systems and in the prediction of overall performance. The possibility of buckling, over-deflection, and wall crushing are evaluated simultaneously by actual tests. No attempt to explain the soil-pipe interaction phenomenon is necessary in the use of this method and the end results leave nothing to be estimated on the basis of judgment.

For example, if tests show that for a given soil compaction at 25 ft of cover, a flexible pipe deflects 3 percent, and in every other way performs well, the actual load on the pipe and the soil modulus are academic. Thus, a pipe installation can be designed with a known factor of safety provided that enough empirical test data are available. In collection of these data, pipe was installed in a manner similar to that used in actual practice and the height of cover increased until performance levels were exceeded. The procedure was repeated many times and a reliable empirical curve of pipe performance versus height of fill was plotted. The use of these empirical curves or data eliminates the need to determine the actual soil pressure since the pipe performance as a function of height of cover is determined directly. Equally good empirical approaches to study of the deflection mechanism are:

The study of actual field installations.

The simulation of a large enough earth cover in a soil test box to exceed the performance limits of the pipe.

To avoid the problem of having to establish design data for the infinite variety of installations and bedding conditions that are found in the field, the following design bases have been chosen:

The embankment condition is selected as critical. (The results are conservative for other than embankment conditions.)

Time lag or settlement of the embankment is included by analyzing long-term values of deflection.

An added advantage of this system is that, by a single test, not only can ring deflection be determined but performance limits such as ring crushing, strain and wall buckling can be noted and analyzed. The use of such data may be considered the most reliable method of design and is recommended when available. Some of the pipe products for which empirical test data have been determined are as follows:

Asbestos-cement pipe (AC)

Corrugated steel pipe

Ductile iron pipe

Fiberglass reinforced plastic pipe (FRP)

Polyethylene pipe (PE)

Polyvinyl chloride pipe (PVC)

Reinforced plastic mortar pipe (RPM)

Steel pipe (CMC-CML)

Substantial data are available for PVC sewer pipe made in accordance with ASTM D3034 with minimum pipe stiffness of 46 lb/in^2 and has been compiled by researchers at the Buried Structures Laboratory, Utah State University. The results of many measurements are categorized in Table 3.5 according to soil type, soil density, and height of cover. Deflections presented in Table 3.5 represent the largest deflections encountered under the conditions specified. Data presented in this manner are designed to provide various options for design engineers. Its use, in most cases, will show that several engineering solutions may be available and economic inputs may suggest a proper solution.

For example, suppose PVC sewer pipe (ASTM D3034 DR 35) with a minimum pipe stiffness of 46 lb/in^2 is to be installed where the native soil is a class IV clay. Ninety percent of the line will be at depths as great as 20 ft. The engineer has selected 7.5 percent deflection as his design limit. According to Table 3.5, the native class IV material could be used for that portion of the pipeline with less than 14 ft of cover if compacted to 75 percent of standard Proctor thereby insuring deflections less than 7.5 percent. However, groundwater conditions may make compaction difficult, even impossible, or may result in subsequent reduction in soil strength. If this is the case, class I, II, or III material may be imported and used with appropriate embedment procedures to limit deflection to 7.5 percent. The choice will be based on availability, convenience, and consequently, on cost. For the deep portion of the line, class III material compacted to 85 percent, class II material compacted to 80 percent, or class I material without compaction could be used successfully.

Pipe Design Criteria

Design methods for installation design have been discussed. However, no design can be effected without performance criteria. Performance criteria are usually established by the design engineer based upon required performance and capabilities of specified products. When a capability of a product is reached or exceeded, it is said that a performance limit has been reached. Each product will exhibit one or more performance limits for each application. Performance limits are established for each product to prevent conditions that may interfere with the design function including the life of the product.

Performance limits

For buried pipes, as for most structures, performance limits are either directly related to stress, strain, deflection, or buckling. It is not im-

TABLE 3.5 Long-Term Deflections of PVC (SDR 35) Pipe (Percent)*†

ASTM embedment material classification‡	Density (Proctor) AASHO T-99, %	Height of cover, ft													
		3	5	8	10	12	14	16	18	20	22	24	26	28	30
Manufactured granular angular (Class I)		0.2	0.3	0.4	0.5	0.6	0.7	0.9	1.0	1.1	1.2	1.3	1.4	1.5	1.6
Clean sand and gravel (Class II)	90	0.2	0.3	0.5	0.7	0.8	0.9	1.1	1.2	1.3	1.4	1.6	1.7	1.8	2.0
	80	0.9	1.4	2.3	3.2	3.6	4.1	5.0	5.5	6.0	6.4	7.3	7.7	8.2	9.1
Sand and gravel with fines (Class III)	90	0.2	0.4	0.6	0.8	0.9	1.1	1.2	1.4	1.6	1.7	1.9	2.1	2.2	2.3
	85	0.7	0.9	1.7	2.2	2.6	3.0	3.5	3.9	4.3	4.8	5.2	5.6	6.0	6.5
	75	1.1	1.8	2.9	3.8	4.5	5.5	6.8	8.5	9.9	11.3	12.7	14.1	15.5	16.8
	65	1.3	2.4	3.6	4.7	5.5	6.8	8.5	9.6	11.4	13.0	14.5	16.0	17.3	18.0
Silt and clay (Class IV)	85	0.65	0.9	1.7	2.2	2.6	3.0	3.5	3.9	4.3	4.8	5.2	5.6	6.0	6.5
	75	1.3	2.3	3.3	4.3	5.0	6.5	7.8	9.5	10.6	12.2	13.5	15.0	16.3	17.0
	65	1.3	2.4	3.6	4.7	5.5	8.0	10.5	12.5	15.0	17.6	20.0	22.0	24.0	26.0

*Test data indicates no length of pipe installed under conditions specified will deflect more than is indicated; the pipe will deflect less than the amount indicated if specified density is obtained.

†Listed deflections are those caused by soil loading only and do not include initial out of roundness, etc.

‡Embedment material classifications are as per ASTM designation D2321 Underground Installation of Flexible Thermoplastic Sewer Pipe.

SOURCE: Data obtained from Utah State University report.

plied that stress, strain, deflection, and buckling are independent, but only convenient parameters on which to focus ones attention. For a particular product, certain performance limits are not considered because others will always occur first. The following is a list of performance limits which are often considered in design and could be thought of as possible responses to soil pressure:

Wall crushing (stress)

Wall buckling

Reversal of curvature (deflection)

Overdeflection

Strain limit

Longitudinal stresses

Shear loadings

Fatigue

Delamination

Wall crushing.   Wall crushing is the terminology used to describe the condition of localized yielding for a ductile material or cracking failure for brittle materials. This performance limit is reached when the in-wall stress reaches the yield stress or the ultimate stress of the pipe material. The ring-compression stress is the primary contributor to this performance limit. (See Fig. 3.14.)

$$\text{Ring compression} = \frac{P_v D}{(2A)} \qquad (3.11)$$

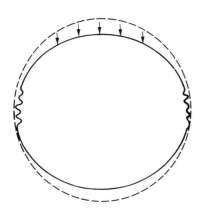

Figure 3.14 Wall crushing at the 3 and 9 o'clock positions.

where P_v = vertical soil pressure
D = diameter
A = cross-sectional area per unit length

However, wall crushing can also be influenced by the bending stress.

$$\text{Bending stress} = \frac{Mt/2}{I} \qquad (3.12)$$

bending moment per unit length
t = wall thickness
I = moment of inertia of wall cross section per unit length

Wall crushing is the primary performance limit or design basis for most "rigid" or brittle pipe products. This performance limit may also be reached for stiffer flexible pipes installed in highly compacted backfill and subjected to very deep cover. A quick check for this performance limit can be made by comparing the ring compression stress with yield and/or ultimate strengths.

Wall buckling. Buckling is not a strength performance limit, but can occur because of insufficient stiffness. The buckling phenomenon may govern design of flexible pipes subjected to internal vacuum, external hydrostatic pressure, or high soil pressures in compacted soil. (See Fig. 3.15.)

The more flexible the conduit, the more unstable the wall structure will be in resisting buckling. For a circular ring in plane stress, subjected to a uniform external pressure, the critical buckling pressure is

$$P_{cr} = \frac{3EI}{R^3} \qquad (3.13)$$

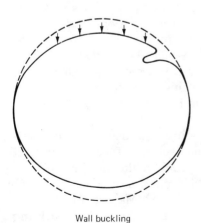

Wall buckling

Figure 3.15   Localized wall buckling.

For a long tube in plane strain, E must be replaced by

$$E = \frac{E}{(1 - \nu^2)}$$

Also I may be replaced by

$$\left(\frac{t^3}{12}\right) \times 1$$

in Eq. (3.13).

$$P_{cr} = \frac{Et^3}{4(1 - \nu^2)R^3} \tag{3.14}$$

For buckling in the inelastic range (materials with pronounced yield points), the critical buckling pressure in terms of the yield point (σ_y) is

$$P_{cr} = \frac{t}{R}\left(\frac{\sigma_y}{1 + 4\sigma_y R^2/Et^2}\right) \tag{3.15}$$

The limiting value of the above equation as the pipe thickness becomes small, is

$$\frac{Et^3}{4R^3}$$

which is less than Eq. (3.14). In fact, in all cases Eq. (3.15) is less than

$$\frac{(\sigma_y t)}{R}$$

or less than the pressure corresponding to the yield point stress. The above equations apply only to a hydrostatic condition, that is, for a conduit completely submerged in a medium that has zero shear strength. The above equations would therefore be valid for checking buckling resistance of a pipeline used for a river crossing, or for a liner pipe, or a pipe in a saturated soil, or a line subjected to an internal vacuum. This analysis does not include initial ellipticity of the conduit.

Most conduits are buried in a soil medium that does offer considerable shear resistance. An exact rigorous solution to the problem of buckling of a cylinder in an elastic medium would call for some advanced mathematics and since the performance of a soil is not very predictable, an exact solution is not warranted. Meyerhof and Baike developed the following formula for computing the critical buckling force in a buried circular conduit:[12]

$$P_{cr} = \frac{2}{R} \sqrt{\frac{KEI}{1 - \nu^2}} \qquad (3.16)$$

If the "subgrade modulus" K is replaced by the soil stiffness E' we have

$$P_{cr} = 2 \sqrt{\frac{E'}{1 - \nu^2}\left(\frac{EI}{R^3}\right)} \qquad (3.17)$$

In both Eqs. (3.16) and (3.17) initial out-of-roundness is neglected but the reduction in P_{cr} because of this is assumed to be no greater than 30 percent. As a result, it is recommended that a safety factor of 2 be used with the above formula in the design of a flexible conduit to resist buckling. The Scandinavians have rewritten the above formula for critical buckling pressure as follows:

$$P_{cr} = 1.15\sqrt{P_b E'} \qquad P_b = \frac{2E}{1 - \nu^2}\left(\frac{t^3}{D}\right) \qquad (3.18)$$

Actual tests show that while the above equations work fairly well for steel pipe, the equations are conservative for either plastic pipe or fiberglass reinforced plastic pipe. However, one of the above equations should be used for design, keeping in mind that predicted buckling pressure will be on the conservative side for most plastic pipe.

Extreme caution should be used when considering large diameter pipes. The above equations assume the external pressure (or internal vacuum) to be essentially constant around the pipe. This condition is not met when a very large pipe is placed in shallow burial below the water table. In this case, the hydrostatic pressure can vary substantially from the top to the bottom of the pipe.

Over-deflection.   Deflection is a design parameter for flexible pipes and so-called "semirigid" or "semiflexible" pipes. It is rarely, if ever, considered in the design of rigid pipe installations.

Flexible pipe products will have a deflection design limit (Fig. 3.16). This design limit is not a performance limit, but is often based on a performance limit with a safety factor. For example, PVC pipes will not start a reversal of curvature until about 30 percent deflection. (See Fig. 3.17.) Thus a design deflection of 7.5 percent is based on a safety factor of 4.

Not all design deflections are based on reversal of curvature. For cement-lined steel and ductile iron pipe, the design deflections are based on deflection limits (performance limits) which produce substantial cracking in the cement lining. Other products have deflection limits to limit bending stresses or strains. The design engineer must

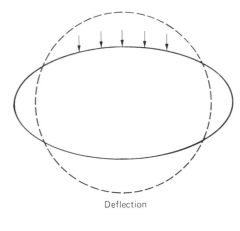

Deflection

Figure 3.16   Ring deflection in a flexible pipe.

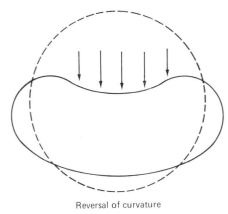

Reversal of curvature

Figure 3.17   Reversal of curvature due to over-deflection.

be aware of each product's limitations for design calculations and in assessing adequate safety.

The "semirigid" and "semiflexible" products depend on their deflection capability to carry the imposed soil load—just as all flexible products do. Thus a deflection consideration must be made in design. For such products, bending stress and bending strain may also become limiting performance criteria. Such products are often cited as having only the positive attributes of both rigid and flexible pipe. However, tests have shown that these same products can and do exhibit the combination of performance limits of both rigid and flexible pipes which makes design analysis more complicated.

The calculated design deflection should always be equal to or less than the design deflection limit for the particular product. The design deflection is calculated by one of the methods described under the flexible pipe analysis section of this chapter. Traditionally, Spangler's

Iowa formula has been used. Finite element methods are starting to be used and will be the method of the near future.

Reversal of curvature.   Reversal of curvature is a deflection phenomenon and will not occur if deflection is controlled. A reverse curvature performance limit for flexible steel pipe was established shortly after publication of the Iowa formula. It was determined that corrugated steel pipe would begin to reverse curvature at a deflection of about 20 percent. Design at that time called for a limit of 5 percent deflection, thus providing a structural safety factor of 4.0. From this early design consideration, an arbitrary design value of 5 percent deflection was selected.

Buried PVC sewer pipe (ASTM D3034, DR 35), when deflecting in response to external loading, may develop recognizable reversal of curvature at a deflection of 30 percent. This level of deflection has been commonly designated as a conservative performance limit for PVC sewer pipe. Research at Utah State University has demonstrated that the load carrying capacity of PVC sewer pipe continues to increase even when deflections increase substantially beyond the point of reversal of curvature. With consideration of this performance characteristic of PVC sewer pipe, engineers generally consider the 7.5 percent deflection limit recommended in ASTM D3034 to provide a very conservative factor of safety against structural failure.

Strain limit.   The strain must be limited in certain pipe materials, such as some fiberglass reinforced pipes. This limit is necessary to prevent strain corrosion. Strain corrosion is an environmental degragation of the pipe material which takes place, in a finite time, only after the pipe wall strain is greater than some threshold strain. Proper design calls for the design strain to be lower than this strain limit with some safety factor.

Strain is related to deflection. Therefore, most manufacturers of such products will propose installation techniques for their particular product which will limit deflection and thus limit the strain. Usually only brittle, composite, or highly filled materials have installation designs which are controlled by strain.

Strain described in this section refers to total circumferential strain, which is made up of bending strain, ring compression strain, hoop strain due to internal pressure, and strain due to Poisson's effect. For gravity sewer pipe, the bending strain is largest and other components may be small in comparison. Research studies indicate that bending strain is given by the following equation:

Bending strain

$$\epsilon_b = 6\left(\frac{t}{D}\right)\left(\frac{\Delta y}{D}\right) \qquad (3.19)$$

Ring compression strain

$$\epsilon_c = \frac{P_v D}{2tE}$$ (3.20)

Hoop strain (due to internal pressure)

$$\epsilon_p = \frac{PD}{2tE}$$ (3.21)

Poisson's circumferential strain

$$\epsilon = -\nu \times (\text{longitudinal strain})$$ (3.22)

where ϵ_b = bending strain
ϵ_c = ring-compression strain
ϵ_p = internal-pressure strain
ϵ = circumferential Poisson's strain
t = wall thickness
D = diameter
Δy = vertical deflection
P_v = vertical soil pressure
E = Young's modulus
p = internal pressure
ν = Poisson's ratio

Longitudinal stresses.   Installation design and construction should be such that longitudinal stresses are minimized. Rigid pipe products and many flexible pipe products are not designed to resist high longitudinal stresses. Longitudinal stresses are produced by

1. Thermal expansion (contraction) (major design consideration in welded steel lines)

2. Longitudinal bending

3. Poisson's effect (due to internal pressure)

Thermal stresses in welded steel lines are often produced by welding the pipe during the high-temperature period in the day. Cooling later can cause extremely high tensile stresses. These stresses can be minimized by providing closure welds during cool temperatures or by the use of expansion joints.

Some of the major causes of longitudinal bending or beam action in a pipeline area are

1. Differential settlement of a manhole or structure to which the pipe is rigidly connected

2. Uneven settlement of pipe bedding or undermining, for example, erosion of the soil below it into a water course or leaky sewer

3. Ground movement associated with tidal water

4. Seasonal rise and fall of soil effected by changes in moisture content (for example, most expansive clays)

5. Nonuniformity of the foundation

6. Tree-root growth pressure

This type of bending frequently occurs when pipes are bent to conform to direction changes. Such bending can cause ring buckling. Reissner has also provided an equation for calculating the radius of curvature that will cause ring buckling as follows:

$$R_b = \frac{D}{1.12t/D} \tag{3.23}$$

Shear loadings. Shear loadings often accompany longitudinal bending. The cause can usually be attributed to nonuniform bending or differential settlement. Forces can be large, highly variable, localized, and may not lend themselves to quantitative analysis with any degree of confidence. For this reason, shear force must be eliminated or minimized by design and proper installation.

Fatigue. The fatigue performance limit may be a necessary consideration in both gravity flow and pressure applications. However, normal operating systems will function in such a manner as not to warrant consideration of fatigue as a performance limit; although some fatigue failures have been reported in forced sewer mains.

Pipe materials will fail at a lower stress if a high number of cyclic stresses are present. Pressure surges due to faulty operating equipment and resulting water hammer may produce cyclic stress and fatigue. Cyclic stresses from traffic loading is usually not a problem except in shallow depths or burial. The design engineer should consult the manufacturer for application where cyclic stresses are the norm.

Delamination. Reinforced and laminated products may experience delamination when subjected to ring deflection. Delamination is caused by radial tension and interlaminar shear. In the design of reinforced products, the radial strength is often neglected and radial reinforcement is omitted. However, the resulting radial strength may be adequate if deflections are controlled. Radial tension is given by

$$\sigma_r = \frac{T}{t(R + y)}$$

$$T = \int_{-c}^{y} \sigma \, da$$

where σ_r = radial tension stress
  t = wall thickness
  R = radius
  y = distance from neutral axis to point in question
  $c = t/2$
  σ = stress in tangential direction as function of position in wall (My/I)
  $da = (dy) \times$ (unit length)

A discussion of radial tension in curved members can be found in most advanced solid mechanics texts.

Delamination may also be caused by chemical action. A prime example is the corrosion of reinforcing steel. When corrosion takes place, corrosion products produce interlaminar pressure which can result in delamination. Reinforcement is usually protected and will not corrode except in case of product misapplication.

Safety factors

The need for selecting a design load that is less than the performance limit load arises mainly from uncertainties. These uncertainties are in service conditions, loads, uniformity in materials, and assumptions made in design. Thus, a reduction factor is needed and is usually referred to as a safety factor or factor of safety.

Rigid pipe.   The safety factor for rigid pipe is usually based on the performance limit of injurious cracking.

W_f = load to cause failure (cracking)

W_w = safe working load

$$W_w = \frac{W_f}{\text{SF}}$$

SF = safety factor

The acceptable safety factor is

$$\text{SF} = 1.5$$

Thus, if a load of 2000 lb/ft will cause cracking, the safe design load should be 2000/1.5 = 1333 lb/ft.

Flexible pipe.   Performance limits for flexible pipes are usually deflection related. Safety factors are then often based on deflection instead of being based on load. For example, if a cement-lined pipe has injurious cracking at 3 percent deflection, a design deflection of 2 percent would be based on a safety factor of 1.5. The design engineer has the responsibility to design the installation (pipe, bending, backfill, and so forth) so that the calculated design deflection does not exceed 2 percent.

Each product will exhibit different performance limits and the factor of safety is usually 1.5 or greater. For flexible products which exhibit only deflection as a performance limit, the design deflection is 7.5 percent and the factor of safety is 4 or greater.

The inexperienced design engineer should consider each possible performance limit, in succession, until the performance limit which occurs at the lowest load or deflection is arrived at. The factor of safety is then based on that performance limit. Literature published by the pipe manufacturer is very helpful in assessing the capabilities and limitations of pipe products.

When a pipe deflects under load, bending strains are induced in the pipe wall. These strains vary linearly through the pipe wall. Somewhere within the wall section (usually about the center) these bending strains will be zero.

Profile-wall pipes are designed and manufactured to minimize the use of material by increasing the section modulus of the pipe wall. "Profile-wall" is a relatively new designation, but the concept is not new. Corrugated steel pipe is truly a profile-wall pipe. Some of the newer plastic products introduced in the last several years are of this type. That is, the plastic is placed primarily at the inside and outside walls or in ribs for higher pipe stiffness. Many of these products have been shown to perform with the profile section acting as a unit as designed. For adequate safety, for any such product, the design should include sufficient plastic between the inner and outer walls and/or between the ribs to carry shear and to ensure that the profile section indeed acts as a unit.

The placement of a rigid-like filler material between walls as a substitute, will impart a brittle-like behavior to the pipe and will interfere with the pipe's ability to deflect without cracking. Such pipes often deflect as a flexible pipe and have a brittle behavior and crack under deflection. Some pipes manufactured in this manner are sometimes referred to as semirigid. This is simply a misnomer. Many solid wall PVC pipe and ductile iron pipes are actually more rigid and still behave as flexible pipes.

Finite Element Methods

Introduction

The finite element analysis (FEA) technique was developed primarily for use in the analysis of complex structural systems. The technique was developed to analyze structural responses to different loading conditions. Through the years, the technique has been extended through mathematical relationships and developed in other areas such as fluid mechanics, thermodynamics, geotechnical engineering, groundwater analysis, aerodynamics, and many other areas of science. The approach has evolved into a rather sophisticated mathematical analysis technique. It has proven to be a very useful tool in research and development as well as in everyday analysis.

One area where the use of FEA has been promoted is in soil-structure interaction mechanics. One- and two-dimensional finite elements can be combined into a global matrix. Each element type may be defined with different stiffness properties. The modeling of the nonlinear stress-strain properties of soil has been accommodated through incremental analysis and an iterative solution scheme. This approach has been widely used in the past for the analysis of earth structures, buried pipes, and earth retaining structures. It has allowed the development and use of some very large and/or complex structures. Various loading conditions, subsurface conditions, and structural properties can be modeled mathematically. This is an advantage over physical testing of such structures. However, the user must be forewarned that the FEA results are only as good as the ability to model the behavior of soil-structure interaction. Also, the finite element method often has to be calibrated by comparing FEA results with results from physical tests. Additional FEA limitations may include inaccurate input data, convergence, and roundoff error.

A study completed at Utah State University in 1985 addressed some of the problems of finite element modeling. The responses of flexible pipes under various loading conditions, compaction conditions, and groundwater conditions were analyzed. In order to calibrate the FEA technique, results from physical tests were used for comparison. The actual modeling of some of the different loading schemes brought out the need for additional development of FEA capabilities that had not been addressed in any previous research efforts. These developments have greatly increased the ability to more accurately model soil-structure interaction, particularly for very flexible pipes.

A computer code SSTIPN was obtained and modified by the Utah

State University researchers. This program has a structure similar to the program SAP originally developed by Wilson.[33] Modifications to SAP to include soil modeling were implemented by Ozawa and Duncan.[18] Further enhancements, including interface elements and improved soil models, were also included (Duncan[1]; Wong and Duncan[34]). SSTIPN was modified to run on a VAX computer at Utah State University in 1982 and has since been significantly enhanced and is now available on a personal computer (PC).

The FEA research and program development that was performed included the addition of nonlinear geometric analysis, and improved iteration scheme, modifications to the soil model to include primary loading, unloading and reloading analysis, and improved output files for pipe response analysis and plotting. Applications of the enhanced model included compaction simulation, initial ovalization of the pipe, unsymmetrical compaction and bedding analysis, and pseudotime effects due to saturation and soil structure collapse.

Laboratory testing for the soil properties was also performed. The testing included grain-size analysis, Atterberg limits, compaction, confined compression, and triaxial testing for stress-strain properties of each soil type. The results of the triaxial testing were used to analyze the pipe response for several soil types on the enhanced version of SSTIPN. Due to the numerous modifications of the code to specifically accommodate pipe analysis, the code is now called PIPE. The PC version is called PIPE5.

Enhancements to the finite element program SSTIPN

Finite element method for stress analysis in solid mechanics is a mathematical technique whereby a continuum is idealized by dividing it into a number of discrete elements. These elements are connected to their adjacent elements at the nodes only.

Special shape functions are used to relate displacements along the element boundaries to the nodal displacements and to specify the displacement compatibility between adjacent elements. Once the continuum has been idealized as shown in Fig. 3.18, an exact structural analysis of the system is performed using the stiffness method of analysis.

Equation (3.24) represents the equilibrium equations, in matrix form, for each node in the idealized system. After applying boundary conditions (identifying nodes with fixed or restricted movement), the system of equations can be solved for the unknown nodal displacements. These displacements can in turn be used to evaluate element stresses and strains.

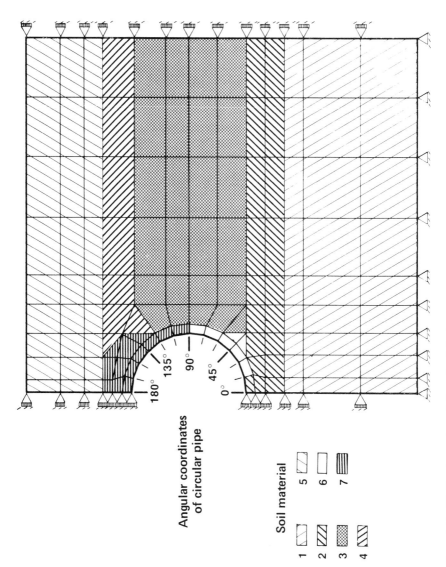

Angular coordinates
of circular pipe

Soil material

1	5
2	6
3	7
4	

Figure 3.18 Finite element mesh for a buried pipe.

$$[K]\{d\} = \{f\} \qquad (3.24)$$

where $[K]$ = global stiffness matrix
    $\{d\}$ = nodal displacement factor
    $\{f\}$ = nodal load vector

The stiffness matrix $[K]$ relates the nodal displacements to nodal forces and is a function of the structural geometry, the element dimensions, the properties of the elements, and the element shape functions.

Finite element analyses for soil-structure interaction problems vary in several ways from finite element analyses for simple linear elastic problems.

1. The soil properties are strain dependent (nonlinear).

2. Different element types must be used to represent the structure (pipe in this case).

3. For flexible pipe the structure may be geometrically nonlinear.

4. It may be necessary in some instances to allow movement between the soil and the walls of the pipe.

The stress strain behavior of the soil is nonlinear; thus the solution procedure must follow the stress condition incrementally. The construction of the soil structure must be followed in steps and the external loads must be added incrementally for the FEA program to follow the nonlinear stress-strain properties of the soil. This particular nonlinear behavior of the soil system has resulted in a special type of analysis that is commonly used in most soil mechanics FEA programs. The basic procedure followed is outlined below. The steps that are described are those that are used in PIPE and PIPE5.

In the finite element analysis of buried pipes, the pipe is modeled using beam elements. These elements are capable of accommodating shear, moment, and thrust. The nodes of the pipe elements are connected to the adjacent soil elements at their common nodal points. Slip between the pipe and soil can be accommodated in the finite element analysis by placing "interface" elements between the pipe nodes and the soil element nodes. These interface elements have essentially no size, but kinematically allow movement between nodes when a specified friction force is exceeded.

Analysis procedure

1. Initial estimates of the stresses and elastic parameters of the soil elements are assumed. Soil properties are nonlinear, and are stress and strain dependent. Due to the soil nonlinearities, the solution pro-

cedure requires modeling that allows for incremental construction of the soil structure and the incremental addition of loads. Initial elastic parameters must be known or assumed to compute the stiffness matrix.

2. An incremental load vector is computed one of two ways. If incremental construction is being modeled, the load vector is computed as the weight of the added soil and/or structure elements for the increment. Alternatively, the load vector may be comprised of external loads resulting from external forces.

3. Incremental nodal displacements are computed for the incremental load vector by solving the system of equations represented by Eq. (3.24).

4. The incremental element strains are computed from a strain-displacement matrix using the nodal displacements. The strain-displacement matrix is based on nodal coordinates of each element and the shape functions used to describe the element behavior. The element strains are then used to compute the element stresses using Hook's law and the initial elastic parameters used in step 1 above. The total stresses, strains, and displacements in the element are computed by adding the incremental stresses, strains, and displacements from the previous increments. An iteration sequence is followed until convergence is achieved. The convergence criterion is that the computed stresses match the initial stresses used to compute the elastic parameters. The total stresses are used to evaluate new elastic parameters for the next loading increment.

5. Once convergence is achieved for a particular load-construction increment, a new incremental load vector is computed and the procedure outlined in steps 2 through 4 is again followed. This method of analysis is called incremental loading method (or equivalent linear method) and is very common to most soil-mechanics finite element analysis programs. The accuracy of the solution is dependent on the assumptions used to derive the stiffness matrix (including the mathematical representation of soil stress-strain response), the size of the loading increment, and many other factors.

The Utah State University research program included the development of a model and its calibration by comparing FEA results with actual physical test data. This FEA research has aided in the enhancement of the computer code. These enhancements have resulted in abilities to better model the actual conditions and predict actual responses.

The computer code PIPE

The computer program PIPE includes soil, beam, bar, and interface elements, and nodal links. The program is structured for a computer that has limited available core storage and thus uses disk storage to store most of the data in up to 16 separate files. For example, the global stiffness matrix, incremental load vector, and displacement vector are each stored in separate files. The structure of the program is such that the individual arrays that are stored in separate files are brought into memory as they are needed in the analysis. Although this may cause the total elapsed time for a particular run to increase due to time needed to retrieve the information on the disk files, the structure of the program makes it easy to adapt for use on microcomputers which have a limited core storage.

Soil model. The soil model that is used is commonly called the Duncan soil model. This soil model assumes that the stress-strain properties of soil can be modeled using a hyperbolic relationship.

Figure 3.19 shows a typical nonlinear stress-strain curve and the hyperbolic transformation that is used. The value of the initial tangent modulus E_t is a function of the confining pressure. Also, the

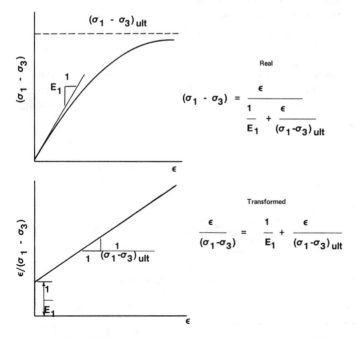

Figure 3.19 Hyperbolic representation of a stress-strain curve. *(After Duncan et al.[1])*

change in the tangent modulus that occurs as strain increases is shown. For a given constant value of confining pressure, the value of the elastic modulus is a function of the percent of mobilized strength of the soil, or the stress level. As the stress level approaches unity (100 percent of the available strength is mobilized), the value of the modulus of elasticity approaches zero. The Mohr-Coulomb strength theory of soil indicates that the strength of the soil is also dependent on confining pressure (see Fig. 3.20). Figure 3.21 shows the logarithmic relationship between the initial tangent modulus and confining pressure. The Duncan soil model combines the variation of initial tangent modulus with confining pressure and the variation of elasticity with stress level to evaluate the tangent modulus of elasticity at any given stress condition. The equation that is used to evaluate the modulus of elasticity as a function of confining pressure strength is

$$E_t = \left[1 - \frac{R_f (\sigma_1 - \sigma_3)(1 - \sin \phi)}{2C \cos \phi + 2\sigma_3 \sin \phi} \right]^2 KP_a \left(\frac{\sigma_3}{P_a} \right)^n$$

where E_t = tangent elastic modulus
P_a = atmospheric pressure used for dimensional purposes
K = elastic modulus constant
n = elastic modulus exponent
σ_1 = major principal stress
σ_3 = minor principal stress (confining pressure)
R_f = failure ratio

Modifications to the Duncan soil model as presented in Duncan[1] use a hyperbolic model for the bulk modulus. The hyperbolic relationship for the bulk modulus is similar to the initial elastic modulus relation-

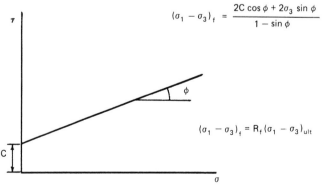

Figure 3.20 Variation of strength with confining pressure. (*After Duncan et al.[1]*)

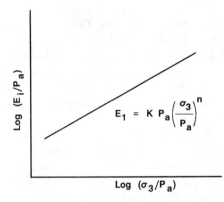

Figure 3.21   Variation of initial tangent modulus with confining pressure. (After Duncan et al.[1])

ship where the bulk modulus is exponentially related to the confining pressure. Figure 3.22 shows the model of the variation of bulk modulus with confining pressure. This particular soil model does not allow for dilatancy of the soil during straining. The equation that is used to relate the bulk modulus to confining pressure is

$$B = K_b P_a \left(\frac{\sigma_3}{P_a}\right)^m$$

where B = bulk modulus
K_b = bulk modulus constant
m = bulk modulus exponent

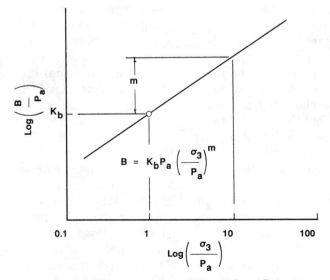

Figure 3.22   Variation of bulk modulus with confining pressure. (After Duncan et al.[1])

The computer code uses the two equations given above to evaluate elasticity parameters that are required in the stiffness matrix. Poisson's ratio and the shear modulus are calculated by the computer using classical theory of elasticity. Limitations are put on the magnitudes of Poisson's ratio in order to remain within the allowable limits of the theory of elasticity. If Poisson's ratio is computed to be more than 0.495, it defaults to 0.495. Likewise if it is computed to be less than 0.0, it again defaults to its lower limit, 0.0.

Shear failure is also tested by evaluating the stress level before the modulus of elasticity is computed. If the stress level is computed to be more than 0.95 of ultimate, the modulus of elasticity is computed based on a stress level of 0.95. This results in a low modulus of elasticity. The bulk modulus is unaffected, thus modeling a high resistance to volumetric compression in shear. A test is also performed to evaluate if tension failure has occurred when computing the elastic parameters. If the confining pressure is negative, the soil element is in tension failure. The elastic parameters are then set to very small values, thus simulating a tension condition. The bulk modulus is set to $0.01B_i$ where B_i is the initial bulk modulus. Poisson's ratio is set to 0.495 and the shear modulus is set to $0.0001B_i$. These constraints appear to be a reasonable approach to modeling soil under shear or tension conditions. The resulting output has been set up to identify the failed elements as the analysis progresses through the incremental loading.

Construction of the stiffness matrix. The stiffness matrix is composed of several parts. In the isoparametric soil elements that are used, the stiffness matrix is recomputed at every iteration. One component is a constitutive matrix relating stress to strain through the elasticity parameters. Another component relates element strains to nodal displacements through the strain-displacement matrix. This matrix is computed based on element types, shape functions, and nodal coordinates. It is not within the scope of this book to derive the above mentioned relationships. The intent is merely to describe how the global stiffness matrix is computed during the analysis.

Beam, bar, and soil elements have their own particular stiffness matrices. A beam element is a three-force element and a bar is a two-force element. Both beam and bar elements are called one-dimensional elements. For these elements, the strain-displacement matrix is derived based on the appropriate shape functions and their cross-sectional area, length, and the angle of inclination of the element. A soil element is a two-dimensional element. It does not transmit moment stresses. The strain-displacement matrix is derived using the x and y coordinates of each node that comprise the element and the shape functions that are used to describe the deformation characteristics of the soil elements.

Small displacement theory. The individual stiffness matrices are computed and stored on separate disk files. Beam and bar stiffnesses are computed only once since their elastic properties are not strain dependent, their shape function matrix is a definite integral, and it is assumed that the nodal coordinates do not change appreciably during the analysis (small displacement theory). Since the soil elements are isoparametric elements, a numerical integration scheme is used to evaluate the strain-displacement matrix at each iteration; however, the nodal coordinates used are the initial coordinates since small displacement theory is used. The elastic portion of the soil stiffness matrix is also recomputed during each iteration since soil elasticity is strain dependent. During an individual iteration, the elastic matrix is evaluated for the soil elements and is combined with the solution of the strain-displacement matrix during the numerical integration. Once the stiffness matrix for the soil elements has been computed, an overall global stiffness matrix is formed by combining stiffness entries from adjacent elements having common nodes. A solution procedure is then followed, as discussed previously, where the nodal displacements are evaluated based on the incremental load vector where the incremental load vector is the nodal force vector due to construction loads or external loads.

Large displacement theory. The derivation of the global stiffness matrix is based on small displacement. However, due to the convenient structuring of the computer code that allows storing individual components of the stiffness matrix into separate disk files, modifications have been made to the code to accommodate large displacement theory.

Execution of PIPE requires the user to prepare a data file that contains all the mesh information and material properties. The data that are required in the input file includes nodal coordinates, element data, structural material and properties, soil material properties (Table 3.6 lists the parameters required for the soil model), construction sequence information, preexisting element stresses, strains, displacements, and external loading information. The data on the input file must be prepared according to specific formats as given in the user's manual.

Preexisting stresses. A convenient feature of PIPE is the specification of preexisting elements. These elements may be soil, structure, or interface elements. The preexisting elements are elements already in

TABLE 3.6   Summary of Required Soil Properties for the Hyperbolic Soil Model

Parameter	Name	Function
K, K_{ur}	Modulus number	Relate E_i and E_{ur} to σ_3
n	Modulus exponent	
c	Cohesion intercept	Relate $(\sigma_1 - \sigma_3)_f$ to σ_3
$\phi, \Delta\phi$	Friction angle parameters	
R_f	Failure ratio	Relate $(\sigma_1 - \sigma_3)_{ult}$ to $(\sigma_1 - \sigma_3)_f$
K_b	Bulk modulus number	Value of B/P_a at $\sigma_3 = P_a$
m	Bulk modulus exponent	Change in B/P_a for tenfold increase in σ_3

SOURCE: After Duncan et al.[1]

place before any construction layers or external loading forces are analyzed. The preexisting elements must have initial stresses specified. Preexisting strains may also be input. For the nodes that are contained in the preexisting elements, any preexisting displacements may also be input. Structural forces may be input for any preexisting structural elements that are in place. Preexisting stresses in the interface elements can also be specified. The *preexisting stress* concept is very convenient when performing a series of analyses. The use of preexisting stresses, strains, and displacements essentially defines the stress condition for the preexisting elements. Construction sequences, therefore, need only be modeled once for a given mesh and soil configuration. The preexisting stresses resulting from that construction simulation can be input for the entire mesh and the subsequent analyses can be performed by adding only combinations of external loads to the mesh. This can save on computer time if the user intends on analyzing the mesh for different loading schemes without repeating the construction sequences.

External loads.   External loads can be input as either concentrated loads or uniform loads. Each loading sequence must have the number of concentrated and uniform loads to be used. Concentrated loads are specified by denoting the node number that will receive the load and the x and y components of the point load. Uniform loads are specified for each element that will receive them. The two nodes of an element with a uniform load are specified along with nodal pressures. The magnitude of the nodal pressure is the acting uniform load. Trapezoidal loading can then be modeled by specifying different magnitudes of the uniform load at each node.

PIPE output.   The results of the analysis of PIPE are contained on a data file specified by the user. The results contain all the input information. Element and node information, material properties, construction and load sequencing, preexisting element information, and initial stresses used for estimating the initial elastic parameters are listed. For each load-construction increment, the user has an option concerning the amount of information that will be contained on the output. If the user does not specify that the results will be printed, the output indicates only the load or construction increment number and the nodal forces that were used in the load vector. If the user specifies that the results are to be printed, the output contains all the information for the nodal load vector, nodal displacements, structural response, soil element strains, and soil element stresses. Nodal displacements include the total displacements for the x, y, and rotation components, and the incremental displacements and rotations for that particular increment.

The structural responses that are listed include the moment, shear, and thrust for each node of each structural member. The listing contains the incremental structural forces and the total structural forces from the accumulated incremental forces.

The soil element strain information includes the soil element strains in x and y direction and the shear strain. Element elastic moduli including elastic modulus, Poisson's ratio, shear modulus, and bulk modulus are also listed for each element. In addition, the principal strains for each element are enumerated.

Soil element stresses that are printed include the horizontal and vertical stresses, shear stresses, and principal stresses. The angle of orientation of the origin of planes with respect to the principal plane, the ratio of major to minor principal stress, and stress levels are also printed out for each element. The stress levels that are printed out indicate the stress condition of each element. If the stress level is greater than 1.0, the element has undergone a local shear failure, and the elastic parameters that were used were based on a stress level of 0.95. If the stress level is between 0.0 and 1.0, the element has not undergone either tension or shear failure and the elasticity parameters that were computed were based on the indicated stress level. If the stress level is listed to be 1.0, the element has undergone a tension failure. The element elasticity parameters that were used for this condition were, as indicated in a previous section, very small, to allow for the displacements that would occur for a soil element in tension.

Enhancements included in PIPE

The program PIPE is specifically designed for flexible buried pipe analysis but can be used for rigid pipe analysis. Many of the changes have improved the cosmetics of the output and have improved the analysis of the pipe response without additional calculations. The changes to the code that are discussed involve inclusion of geometric nonlinear analysis; improvement of the soil model to include primary loading, unloading, and reloading; an expanded procedure to increase the number of iterations to convergence; a modified method to improve the stability of the solution on unloading and reloading; and improved output for easier analysis of the pipe response and plotting, and subsequent analysis for preexisting stresses.

Geometric nonlinear analysis.   Most finite element programs have been developed based on small-strain theory. In small-strain theory, it is assumed that during the FEA analysis, the resulting element strains and nodal displacements are too small to justify reevaluation of the stiffness matrix components that were derived based on the nodal coordinates. The stiffness matrix component that relies on the nodal coordinates is the strain-displacement matrix that is derived from the shape-function matrix. The isoparametric element that is used is one in which the shape-function matrix is evaluated at every iteration of the analysis by a numerical integration scheme. Its evaluation is partly based on the x and y coordinates of the element's nodes that define the size and shape of the element.

The strain-displacement matrix, as previously defined, relates the strains that occur in each element based on the displacements of each of the element nodes. In the bilinear element, the shape functions are linear, which result in constant magnitude of strain across each element. However, the magnitudes at strain vary from element to element. Thus, one could visualize a three-dimensional surface showing the x, y, or shear strain across each element and having discontinuous magnitudes of the element boundaries. Use of higher order shape functions would result in a three-dimensional surface with a higher degree of continuity at the boundaries for each increase in degree of the shape function.

In the incremental loading procedure, the nodal coordinates are established at the initial execution of the program. Construction is modeled by adding rows of elements and solving for nodal displacements due to the weights of the newly added elements. However, as the construction sequence is followed, the displaced nodes are not recognized in the stiffness matrix. The assumption of small-strain theory has been investigated in other works and has been shown to provide acceptable results, especially in view of the other inaccuracies in the analysis.

The finite element analysis, which does evaluate the stiffness matrix based on deformed nodal coordinates, is defined as a geometric nonlinear analysis. Thus, one which includes both nonlinear stress-strain properties and large displacement theory performs material and geometric nonlinear analysis.

There has been some concern that the small-strain theory that has been used in the FEA of flexible pipes was inducing some inaccuracies in the results. The addition of the geometric nonlinear analysis has been included by modifying the nodal coordinates of each node at the end of each loading increment. The elemental stiffness matrices of all elements need to be reevaluated at every loading increment due to the changing nodal coordinates. The stiffness matrices of the structural elements are developed partly on the basis of the element length and its inclination. Thus, the structural element stiffness matrix components are reevaluated based on the nodal deformations. Since the soil stiffness matrices are reevaluated at each iteration due to changing material elastic properties, an additional step to reevaluate the strain-displacement matrix (using deformed coordinates) is necessary.

The geometric nonlinear analysis has been used to help determine initial deflections by means of compaction simulation. Also, modifications have given the program the ability to model internal pressure loads and rerounding effects with incremental loading.

Enhanced soil model. The Duncan soil model, as described in a previous section, was developed to model deformation characteristics of soil as the confining pressure of the soil increases. Duncan et al.[1] gave a brief account of the behavior of soil on unloading and reloading. The Duncan soil model could accommodate unloading and reloading by identifying the elastic modulus constant K (defined previously), as the unloading and reloading modulus. A typical stress-strain curve of soil which has undergone primary loading, unloading, and reloading is shown on Fig. 3.23. It can be seen that the soil does not unload to a

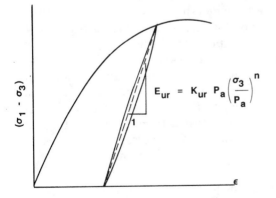

$$E_{ur} = K_{ur} P_a \left(\frac{\sigma_3}{P_a} \right)^n$$

Figure 3.23 Unloading-reloading modulus. *(After Duncan et al.[1])*

zero strain as the stress decreases, and that the unloading tangent modulus of elasticity (slope of the unloading stress-strain curve) is much higher than the slope of the primary loading curve. Duncan et al.[1] indicated that the unloading modulus is independent of stress level. Thus, the slope of the unloading stress-strain curve will not change if unloading is performed at any point on the primary stress-strain curve. They also indicate that the unloading modulus is dependent only on confining pressure and the bulk modulus is not a function of the stress history of the soil.

The equation that relates the unloading-reloading modulus to other soil properties is

$$E_{ur} = K_{ur} P_a \left(\frac{\sigma_3}{P_a}\right)^n$$

where K_{ur} is the unloading reloading constant and E_{ur} is the unloading reloading modulus.

In the original Duncan soil model, if one wanted to use the unloading modulus, the elastic modulus constant that was used for the soil parameters was the unloading constant, K_{ur}. No mechanism was given to evaluate the stress history of the soil elements and the appropriate modulus if unloading is detected. Also, it was not possible to change the soil elasticity parameters during the analysis in order to simulate loading and unloading all within a single analysis. Additionally, it has been noted that some soil elements will not respond to unloading when a given external loading pattern causes a decrease in stress.

Determining the maximum stress. The most desirable condition is to provide a means of monitoring the stress history of each soil element and to use an unloading modulus when the soil stresses are detected to be less than a maximum previous stress. An improved soil model was developed which includes both primary loading parameters and unloading-reloading parameters. The stress condition of a soil element is uniquely determined by the values of the maximum and minimum principal stresses. For plane-strain analysis, the intermediate principal stress is assumed to be equal to the minimum principal stress. Several different schemes have been tested to monitor the stress history of each soil element: maximum deviator stress, maximum confining pressure, maximum principal stress, and maximum average stress.

The schemes were investigated in view of Mohr's circle analysis. These investigations show that the best variable for testing the stress condition of the soil elements is the average stress (or the center of

Mohr's circle). The center of Mohr's circle gives a general indication of the stress condition, dependent on both maximum and minimum principal stresses. If the position of the center of the circle is decreasing, an unloading modulus is in effect. The unloading-reloading modulus is also in effect until the position of the center of the circle exceeds a maximum position indicated by the stress history.

The average principal stress is monitored for each element and compared to its maximum average stress. The soil model uses an unloading modulus if the average stress is less than the maximum. A mechanism is provided to simulate maximum past pressures by inputting values for maximum stresses for each soil element, similar to the pre-existing stress concept.

Behavior of other soil parameters. The discussion given by Duncan et al. (1980) indicates that the only soil parameter that is a function of stress history is the elastic modulus constant. However, there appears to be an insufficient data base to substantiate these remarks. Poisson's ratio v, is computed based on the bulk modulus B and elastic modulus E by

$$v = \frac{3B - E}{6B}$$

Duncan's recommendation is that the modulus of elasticity is from 1.2 to 3.0 times greater on unloading than on primary loading depending on the soil density. If the bulk modulus is invariant of stress history, the value of Poisson's ratio will become very small if the modulus of elasticity increases by a factor of 2 or 3. This would indicate that a soil will have very little lateral deformation with changing vertical stress if the soil has seen a stress condition greater than the existing stresses.

The behavior of the soil parameters on primary loading and unloading was investigated using triaxial soil tests at Utah State University. The results of this testing program indicate that the bulk modulus behavior is very unpredictable on loading and reloading. It is difficult to make any definite observations on the behavior of the bulk modulus. However, the elastic modulus exponent, in some cases, is dependent on stress history. Consequently, the soil model has been modified to use both the unloading elastic modulus constant and the unloading elastic modulus exponent.

Magnitude of unloading modulus constant. As mentioned, Duncan et al.[1] recommend that the unloading modulus constant be approximately 1.2 times higher than the primary loading constant for stiff soils and 3.0 times higher for soft soils. These approximate factors appear to work relatively well, in view of the results of the triaxial testing program. In fact, the modulus constant has been as much as four times higher on unloading than on primary loading. This leads to the phenomena of small or even negative values of Poisson's ratio.

Iteration procedure

The iteration procedure accommodates the changes in elastic moduli when they occur. The soil elements are monitored to test whether they are on the primary loading curve or on the unloading-reloading curve during the first iteration of the previous loading increment. If the results of the first iteration indicate that the soil element is changing from one curve to another, the element condition is flagged and the second iteration follows the same logic as the first, except that the correct modulus is used to evaluate the elastic parameters based on the stresses from the final iteration of the previous loading increment. The third iteration that follows uses the average stresses to compute new element properties and responses to the current loading increment. It appears that at least three iterations are required if soil unloading-reloading is to be included. However, since the results of the analysis reflect an equilibrium condition, neighboring elements to those that changed their stress condition at the first iteration may not have come to "equilibrium" at the end of the third iteration, particularly if the resulting stresses of the second (or later) iteration indicate that an element should change from one soil model to another. Changing soil models is only permitted on the first iteration. This may cause some difficulties in the strain compatibilities of the solution.

One of the inputs of the data file for PIPE includes a variable for the desired number of iterations. All analyses that have been subsequently performed using the soil unloading-reloading model have used four iterations. A sensitivity study has been performed to evaluate the number of iterations to be used. It appears that four iterations is the optimum when unloading-reloading is included.

PIPE output

The goal was to make the program more user friendly with respect to easier analysis of the PIPE response. The elimination of unnecessary output, the preparation of results for plotting, and the structuring of data files so that calculated stresses can be treated as preexisting stresses for a subsequent analysis, are program enhancements that have been made. Also, computer graphicas have been incorporated to help visulize the modeling process. Figures 3.24 through 3.27 are computer generated displays produced by PIPE5.

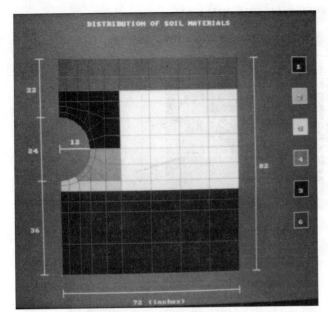

Figure 3.24   Photograph of PC monitor display showing various soil types and/or compactions used in an FEA model.

Printed results.   The output of SSTIPN, as discussed previously, consists of a single output file which contains all the results of a given analysis. So that the user can examine the results, the voluminous output must be printed and the results of each loading increment that was printed must be examined. This procedure can be quite cumbersome, especially when making production runs where only a few variables are needed to present the results. Additionally, the structural response is printed in terms of nodal forces (shears, moments, and thrusts) for each structural element. For example, the design criteria for the FRP pipe are pipe-wall strains and thus, the user must compute strains based on the nodal forces. The output of PIPE is such that computed strains due to thrust and bending are printed. Ring deflections are also printed in terms of percent vertical and percent horizontal deflection for the pipe. Thus, the printed output can easily be examined to evaluate the pipe response. The user may still wish to examine the other parameters, which are still included.

Also incorporated is a data check sequence where the element information is processed to test if the data have been input correctly. Element areas are computed based on the element nodes. If the nodes are not input in a counterclockwise manner, the element area is computed as a negative area, and the user can then identify input errors on the element data more easily. Soil elements which have been evaluated on

DISTRIBUTION OF SOIL MATERIALS

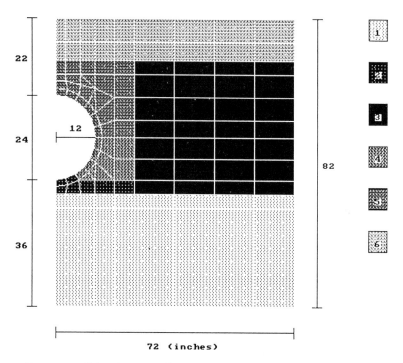

Figure 3.25   Printed output showing various soil types and/or compactions used in an FEA model.

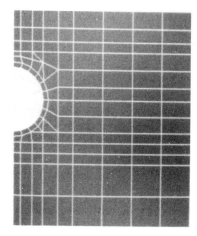

Figure 3.26 Photograph of PC monitor display of the FEA mesh.

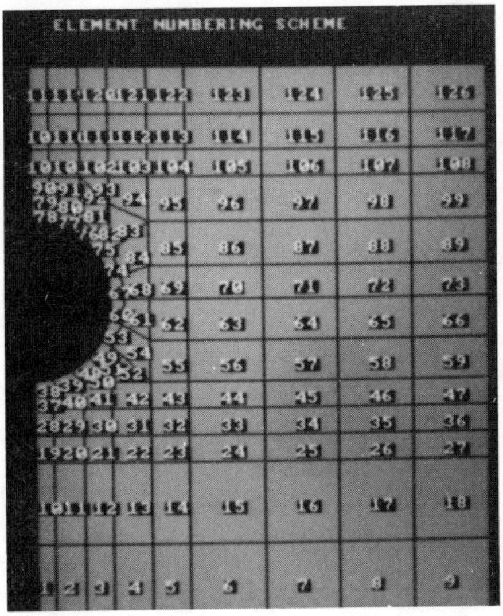

Figure 3.27 Photograph of PC monitor display of an FEA mesh with element numbers.

the unloading model are identified by their stress level. Unloading and rebound elements have a negative stress level.

Additional output.   There are several additional output files that have been included in PIPE to accommodate data processing. For each loading increment, the user has the option of having the results printed on separate files. The option includes having the stresses, strains, and displacements printed to separate output files. This allows the user to use the stresses, strains, and displacements as preexisting stresses for any subsequent runs.

In addition to having an option to printing particular results to separate output files, an option is included to have the ring deflections separately printed to an output file. This option exists for every load increment. Combinations of having ring deflection files, and/or stress, strain, and displacement files are included. Results of a given run can be easily viewed by examining the load deflection curve, and therefore, viewing the ring deflection file facilitates a much faster review of the results.

Plotting.   Several output files have been created that are compatible with the plotting routines. The mesh information is stored on a separate file that has a compatible format with mesh plotting routines. Pipe strains are also printed out to a file that is used to plot the strains versus position of the pipe. The ring deflection file previously described is also used to plot the load deflection curve for a given analysis. Thus, the results of a given run can be analyzed through the out-

Figure 3.28 Small test cell.

Figure 3.29 Large test cell.

put files, and presented graphically through plotting files that are used by the post-processor plotting programs.

Example applications

Some results from applications of the FEA program PIPE are included here and are compared with measured responses from actual tests conducted in soil load cells at Utah State University (see Figs. 3.28 and 3.29). The comparisons that are shown are for pipe with a 10-lb/in² pipe stiffness. Test-cell soil compaction conditions that are included for comparisons are

1. Ninety percent relative compaction with homogeneous conditions
2. Ninety percent relative compaction with poor haunches
3. Eighty percent relative compaction with homogeneous conditions

Soil parameters used in the FEA program are listed in Table 3.7.

In the buried pipe tests, every attempt was made to achieve homogeneous conditions when called for. However, the flexible nature of the pipe does not always allow for a high uniform compaction in the haunches and around the shoulders and crown of the pipe. Therefore, homogeneous conditions that are attempted in the test cell or for that matter in an actual installation will result in some variation in density. Of course the FEA program can perfectly model the homogeneous soil condition. When a test pipe was installed in the soil box with poor haunches, no attempt was made to compact the soil in the haunch area. Finite element modeling of homogeneous and poor haunch conditions is well defined because numerically all soil elements in each homogeneous condition have identical stress-strain properties.

Comparisons of the FEA results with those of the soil-box tests can be made using pipe-strain and load-deflection results. For the pipe-strain plots, tension bending strains on the outside fibers are considered positive. Thrust strains around the circumference of the pipe are also included. The load-deflection plots show the vertical and horizontal ring

TABLE 3.7   Soil Parameters for Silty Sand

Relative compaction standard, %	Density, lb/in³	ϕ, deg	Δ, deg	c, lb/in²	K	n	R_f	K_b	m	K_o	K_{ur}	n_{ur}
90	0.065	30	0.	8.3	480	0.44	0.75	80	0.38	0.48	720	0.44
80	0.058	30	0.	3.5	350	0.28	0.89	15	0.40	0.37	525	0.28

NOTE: ϕ, friction angle; Δ, friction angle reduction for 10-fold increase in lateral pressure; C, cohesion intercept; K, elastic modulus constant; n, elastic modulus exponent; R_f, failure ratio; K_b, bulk modulus constant; m, bulk modulus exponent; K_0, earth pressure coefficient; K_{ur}, unload/reload modulus constant; N_{ur}, unload/reload modulus exponent.

deflections in terms of surcharge pressure. The zero point for the load-deflection plots for the load-cell tests is referenced to the deformed state of the pipe after compaction. In the FEA plots, the zero reference for ring deflection is based on the initial undeformed condition. Thus, in the load-deflection illustrations, the zero point of deflection should be considered when direct comparisons are made between results from the FEA and the soil test cell. Plots of pipe-wall strain for the soil test cell and for the FEA results are both referenced from the same unstrained condition. These plots show bending and thrust strain versus position on the pipe. The zero-degree position on the pipe is at the invert, the 90-degree position is at the springline, and the 180-degree position is at the crown as shown in Fig. 3.18. The values for pipe strain from 180 to 360 degrees are symmetric with 0 to 180 degrees for the FEA because the FEA mesh presented here used an axis of symmetry for the analysis of symmetric bedding.

Homogeneous installation at 90 percent relative compaction. Figures 3.30 and 3.31 show the soil-box test results for a 10-lb/in^2 pipe installed with homogeneous compaction at 90 percent of standard proctor maximum dry density. Physical pipe data are as follows:

Parameter	Curve	
	A	B
Stiffness, lb/in^2	10	10
Thickness, in	0.285	0.300
Surface pressure, lb/in^2	48.9	50.0
Vertical deflection, %	5.53	4.82
Horizontal deflection, %	3.74	2.52

Figure 3.30 shows the load-deflection curve and Fig. 3.31 shows pipe strain versus position on the pipe for a surcharge pressure of 48.9 lb/in^2. Features of these results to note are the shape of the load-deflection curve, relative magnitudes between the horizontal and vertical ring deflections, and shape and magnitudes of bending and thrust strain. This condition was modeled with FEA in several ways. These illustrations also show the results from the FEA for a homogeneous 90-percent relative compaction with no compaction simulation. There is a marked similarity between FEA and test data. The pipe-strain plot in Fig. 3.31 indicates that the magnitudes of pipe strain at a surface pressure of 50.0 lb/in^2 are fairly comparable. The ring deflections determined from experiment and for FEA also compare quite closely.

Figures 3.32 and 3.33 show the results of the FEA for the homogeneous dense condition including compaction simulation during construction. The physical pipe data are as follows:

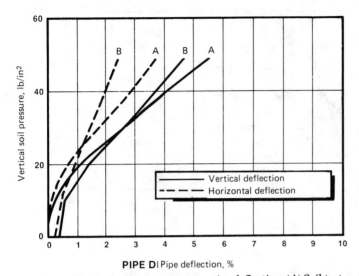

Figure 3.30 Vertical soil pressure versus pipe deflection. (A) Soil test cell data, 90 percent relative compaction; (B) FEA, no compaction simulation.

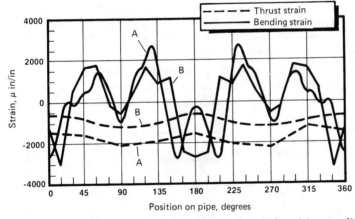

Figure 3.31 Pipe strain as function of circumferential position, conditions as in Fig. 3.30.

	Curve	
Parameter	A	B
Stiffness, lb/in^2	10	10
Thickness, in	0.285	0.300
Surface pressure, lb/in^2	48.9	50.0
Vertical deflection, %	5.53	5.42
Horizontal deflection, %	3.74	3.14

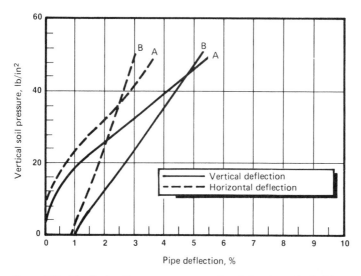

Figure 3.32 Vertical soil pressure versus pipe deflection. (*A*) Soil-box data, 90 percent relative compaction, silty sand; (*B*) FEA with compaction simulation.

The compaction simulation load-deflection curve in Fig. 3.32 lost some of the initial steepness as compared with Fig. 3.30. However, the difference between vertical and horizontal deflection is maintained. Deflections of Fig. 3.32 are similar in magnitude to those of Fig. 3.30. Figure 3.33 shows pipe-strain plots for compaction simulation and a surface pressure of 50.0 lb/in². A comparison of data in Fig. 3.33 with

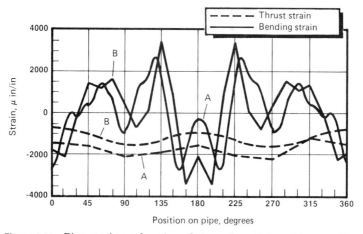

Figure 3.33 Pipe strain as function of circumferential position, conditions as in Fig. 3.32.

data in Fig. 3.31 shows that compaction simulation did improve the correlation between FEA and test results. The general shape, maxima, and magnitudes all compare very well.

Additional comparisons that were made with this condition included soft elements in the shoulder areas of the pipe. Because soil placement techniques do not allow compaction directly above the pipe, a completely homogeneous compaction is not obtained in an actual installation. For a flexible pipe the soil will be of a lesser density at the shoulders and crown of the pipe. One noticeable result with soft-crown analyses is that generally the pipe strain at the 135-degree position of the pipe (see Fig. 3.33) increased. This is due to the lowered stiffness of the soil in the shoulders, which allows for more bending deformation in the pipe. Compaction simulation for the soft-crown condition did decrease the bending strains and ring deflections because the soil would respond in the rebound range initially, thus inhibiting deformation at the low pressure ranges. Because compaction simulation did not include adding loads directly over the pipe at the first construction increment, a soft-crown condition was actually created with the homogeneous case. This is because the soil at the crown was uncompacted and did not respond on the stiffer rebound modulus at the lower pressure ranges as did the surrounding soil elements that had received the compaction loads directly.

Poor haunch installation at 90 percent relative compaction. Figures 3.34 and 3.35 show the results for the poor haunch installation with a silty sand soil. A poor haunch condition is, as used here, where soil is placed in the haunch areas but is not compacted. The physical pipe data are as follows:

| | Curve | |
Parameter	A	B
Stiffness, lb/in^2	10	10
Thickness, in	0.285	0.300
Surface pressure, lb/in^2	35.5	30.0
Vertical deflection, %	3.14	2.21
Horizontal deflection, %	1.30	1.09

Figure 3.34 shows the load-deflection response and Fig. 3.35 shows the pipe strain around the pipe for a surface pressure of 35.5 lb/in^2. Again, the initial steepness of the load-deflection curve, the relative magnitudes between the vertical and horizontal deflections, and the shape and magnitude of the strain plots should be noted. The bending strains are higher than before at the 30- to 45-degree positions of the pipe because of the lack of support in the haunch area. Also, a com-

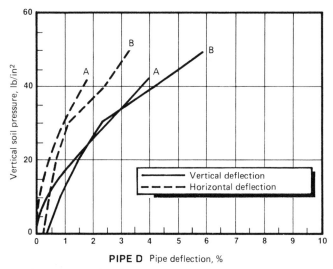

Figure 3.34 Vertical soil pressure versus pipe deflection. *(A)* Soil-box data, 90 percent relative compaction, silty sand, and poor haunch support; *(B)* FEA, no compaction simulation and poor haunch support.

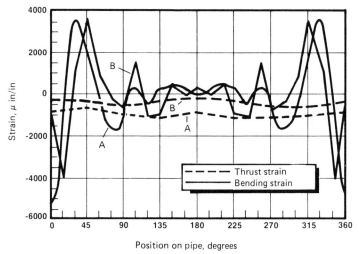

Figure 3.35 Pipe strain as function of circumferential position, conditions as in Fig. 3.34.

parison between the homogeneous installation and the poor haunch installation (Fig. 3.33 and 3.35, respectively), shows noticeable differences in the pipe-strain plots from soil-box tests.

Figures 3.33 and 3.35 also show the FEA results for the poor haunch condition without compaction simulation. The load-deflection plots

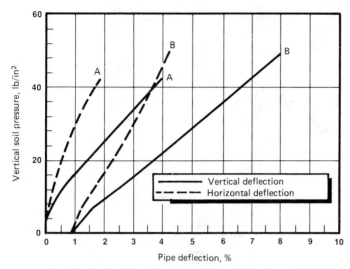

Figure 3.36 Vertical soil pressure versus pipe deflection. (*A*) Soil-box data, 90 percent relative compaction, silty sand, and poor haunch support; (*B*) FEA with compaction simulation and poor haunch support.

show similar behavior, yet the deformations are larger in the FEA results. The pipe-strain plots show very similar peaks of large strain at the 45-degree position and low strains from the springline to crown.

Figures 3.36 and 3.37 show the FEA results for poor haunches with compaction simulation. In the load-deflection plot, the FEA indicates larger deflections. The pipe-strain plots show larger strains in the pipe from the springline to the crown. However, the strain at the invert of the pipe with compaction simulation compared better with measured results. That is, FEA with compaction simulation seems to give a more accurate prediction of strain at the pipe invert as compared with FEA without compaction simulation. The physical pipe data for Figs. 3.36 and 3.37 are as follows:

	Curve	
Parameter	*A*	*B*
Stiffness, lb/in^2	10	10
Thickness, in	0.285	0.300
Surface pressure, lb/in^2	35.5	30.0
Vertical deflection, %	3.14	5.14
Horizontal deflection, %	1.30	2.92

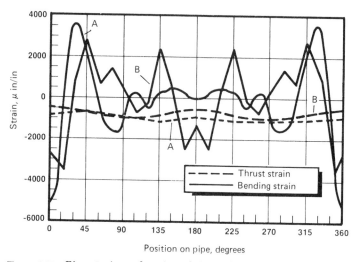

Figure 3.37   Pipe strain as function of circumferential position, conditions as in Fig. 3.36.

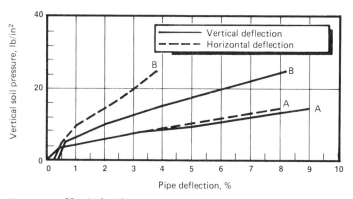

Figure 3.38   Vertical soil pressure versus pipe deflection. (*A*) Soil-box data, 80 percent relative compaction; (*B*) FEA, no compaction simulation.

Homogeneous installation with 80 percent relative compaction.   Figures 3.38 and 3.39 show the test results for an 80-percent relative compaction homogeneous installation. The physical pipe data are as follows:

Parameter	Curve	
	A	*B*
Stiffness, lb/in^2	10	10
Thickness, in	0.285	0.300
Surface pressure, lb/in^2	14.6	15.0
Vertical deflection, %	8.78	3.85
Horizontal deflection, %	7.87	2.06

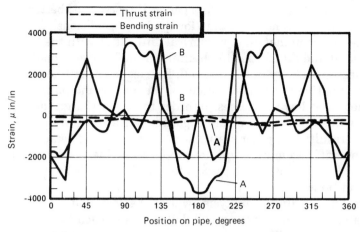

Figure 3.39 Pipe strain as function of circumferential position, conditions as in Fig. 3.38.

The vertical and horizontal deflections are very similar throughout the test, which indicates elliptical deformation as shown in Fig. 3.38. Figures 3.38 and 3.39 also show the results from the FEA for the 80-percent relative compaction homogeneous condition. Although the load-deflection curves show much more deformation with the loose material than with the dense material, the actual comparison of soil-box tests with FEA tests shows that the FEA does not compare quite as well for loose soil conditions. The pipe-strain plots shown in Fig. 3.39 also indicate a generally poorer correlation. In terms of magnitude of the maximum strain, there is some correlation but the overall shape of the pipe-strain plot does not match the measured values as well as for the cases with 90-percent density.

Discussion of results. The incorporation of the compaction simulation for comparison of the response of the FRP pipe improved the comparison for the homogeneous condition for most cases that were attempted. For the nonhomogeneous installation conditions, the compaction simulation did not improve the correlation of FEA and test results. It is possible that nonhomogeneous conditions dominate the response, masking the compaction simulation response. This could be due to the nature of the compaction simulation sequence. Had the compaction sequence individually modeled the backfill condition (poor haunch, soft top, and so forth), the results might have improved. For most cases, the compaction simulation does not improve the results enough to justify the additional computational effort required.

The FEA data and experimental data generally correlated better for the dense installation conditions than for the loose conditions. This is probably due to a combination of numerical difficulties with the finite element method and difficulties in obtaining a uniform soil condition for low to medium density in the test cell. Entries in the stiffness matrix become sensitive to the magnitudes of the elastic and bulk modulus parameters at low stiffnesses. In order to achieve larger deflections, lower values of the bulk modulus parameters are required. This, however, can result in singular matrix warnings, which indicates that entries in the stiffness matrix will not produce reliable results. More work is needed in this area with respect to modeling soil behavior under loose conditions.

The geometric nonlinear analysis (where the formulation of the stiffness matrix accounts for the nodal deflections at each loading increment) does not significantly change the results for installation-condition modeling. The inclusion of the geometric nonlinear analysis would generally predict somewhat higher deflections. For example, an analysis that did not include geometric nonlinearities might predict a vertical ring deflection of 4 percent. The same conditions including geometric nonlinearities would predict ring deflections of around 5 percent. However, for the other types of loading conditions (for instance, rerounding), the formulation of the stiffness matrix must reflect the shape of the pipe.

Summary and conclusions

Good correlation of finite element modeling of flexible pipes with test data requires modeling capabilities not readily available in most existing computer programs. Such capabilities include analysis of stress history of the soil elements to determine whether each element is in primary loading or in unloading and reloading, modification of the iteration scheme to better model the soil response when changing from one stress condition to another, and large deflection theory by modifying nodal coordinates after each load increment. Additionally, postprocessing plotting routines are needed to graphically analyze the pipe response to each loading condition.

The development of these features has allowed for analysis not only of rigid pipe but of flexible pipe with compaction simulation, surcharge pressures, rerounding caused by internal pressurization, and various installation conditions. The results of the analysis of the various installation conditions have shown the effects of shoulder and haunch support on the pipe and suggest that these conditions be considered in pipe and installation design.

The results of this FEA development program at Utah State University have improved the modeling capability of flexible pipe systems. Moreover, an improved understanding of the behavior of the buried flexible pipe has been developed due to the ability to model various installation conditions. The results of the overall study, including the four soil types and various loading conditions, have shown a very good correlation between the FEA results and the measured responses from the physical model tests. This has given strong justification for the use of the finite element method to adequately model various installation conditions, soil materials, loading conditions, pipe sizes, and so forth, without the additional expense of performing extensive physical tests. However, calibration of the FEA model required the results from physical tests. Finite element analysis along with experiments have resulted in a better analytical tool for the evaluation of buried pipe performance. This tool is now available and is being used primarily for research and analysis. It is the design tool of the future.

Bibliography

1. Duncan, J. M., P. Byrne, K. S. Wong, and P. Mabry: "Strength, Stress-Strain and Bulk Modulus Parameters for Finite Element Analysis of Stresses and Movements in Soil Masses," Report no. UCB/GT/80-01, Office of Research Services, University of California, Berkeley, 1980.
2. Dunn, I. S., L. R. Anderson, and F. W. Kiefer: *Fundamentals of Geotechnical Analysis*, J. Wiley & Sons, Inc., New York, 1980.
3. Gere, J. M., and W. Weaver: *Analysis of Framed Structures*, 2d ed., Van Nostrand-Reinhold Company, New York, 1980.
4. Hild, J. W.: "Compacted Fill," in H. F. Winterkorn and H. Y. Fang (eds.), *Foundation Engineering Handbook*, Van Nostrand-Reinhold Company, New York, 1975.
5. Howard, Amster K.: "Modulus of Soil Reaction (E[SA]) Values for Buried Flexible Pipe," *Journal of the Geotechnical Engineering Division*, ASCE, vol. 103, no. GT, Proceedings Paper 127000, January 1977.
6. Janson, Lars-Eric, and Jan Molin: "Design and Installation of Underground Plastic Sewer Pipes," *Proceedings of the International Conference on Underground Plastic Pipe*, ASCE, New York, 1981, pp. 79–88.
7. Jensen, Brent M.: "Investigations of Strain Limits Proposed for Use in Designing PVC Pipe Subjected to External Soil Pressure," M.S. thesis, Utah State University, Logan, Utah, 1977.
8. Katona, M. G., P. D. Vittes, C. H. Lee, and H. T. Ho: "CANDE–1980: Box Culverts and Soil Models," National Technical Information Service, Springfield, Va., 1981.
9. Knight, G. K., and A. P. Moser: "The Structural Response of Fiberglass Reinforced Plastic Pipe Under Earth Loadings," Buried Structures Laboratory, Utah State University, Logan, Utah, 1983.
10. Konder, R. L., and J. S. Zelasko: "A Hyperbolic Stress-Strain Formulation of Sands," *Proceedings of the Second Pan American Conference on Soil Mechanics and Foundation Engineering*, vol. 1, 1963, p 209.
11. Kulhawy, F. H., J. M. Duncan, and H. B. Seed: "Finite Element Analysis of Stresses and Movements in Embankments During Construction," Report no. TE-69-4. Office of Research Services, University of California, Berkeley, California, 1969.
12. Meyerhof, G.G. and L. D. Baike: "Strength of Steel Culverts Sheets Bearing against Compacted Sand Backfill," *Highway Research Board Proceedings*, vol. 30, 1963.

13. Moser, A. P.: "Can Plastic Sewer Pipe Be Installed with 100% Confidence?" Presentation at the 68th Annual Meeting of the ASSE, New Orleans, La., 1974.
14. Moser, A. P.: "Strain as a Design Basis for PVC Pipe?" *Proceedings of the International Conference on Underground Plastic Pipe*, ASCE, New York, 1981, pp. 89–103.
15. Moser, A. P., R. R. Bishop, O. K. Shupe, and D. R. Bair: "Deflection and Strains in Buried FRP Pipe Subjected to Various Installation Conditions," Presented at 64th Annual TRB Meeting, Washington, D.C., 1985, Published in TRB *Transportation Research Record 1008*, Washington, D.C., 1985.
16. Moser, A. P., R. K. Watkins, and O. K. Shupe: "Design and Performance of PVC Pipes Subjected to External Soil Pressure," Buried Structures Laboratory, Utah State University, Logan, Utah, 1976.
17. Nyby, D. W.: "Finite Element Analysis of Soil Sheet Pipe Interaction," Ph.D. dissertation. Department of Civil and Environmental Engineering, Utah State University, Logan, Utah, 1981.
18. Ozawa, Y., and J. M. Duncan: "ISBILD: A Computer Program for Analysis of Static Stresses and Movements in Embankments," Report no. TE-73-4, Office of Research Services, University of California, Berkeley, 1973.
19. Piping Systems Institute: *Course Notebook*, Utah State University, Logan, Utah, 1980.
20. Reissner, E.: "On Final Bending of Pressurized Tubes," *Journal of Applied Mechanics (Transactions of ASME)*, 1959, pp. 386–392.
21. Sharp, Kevan, L. R. Anderson, A. P. Moser, and R. R. Bishop: "Finite Element Analysis Applied to the Response of Buried FRP Pipe Due to Installation Conditions," Paper presented at the 64th Annual TRB Meeting, Washington, D.C., 1985, *Transportation Research Record 1008*, Washington, D.C., 1985.
22. Spangler, M. G.: "The Structural Design of Flexible Pipe Culverts," Bulletin 153, Iowa Engineering Experiment Station, Ames, Iowa, 1941.
23. Timoshenko, S. P., and D. H. Young: *Elements of Strength of Materials*, 4th ed., D. Van Nostrand Company, Inc., Princeton, N.J., 1962, pp. 111, 139.
24. Timoshenko, S. P.: *Theory of Elastic Stability*, 2d ed., McGraw-Hill Book Company, New York, 1961.
25. Timoshenko, S. P.: "Strength of Materials," Part II, "Advanced Theory and Problems," D. Van Nostrand Company, Inc., Princeton, N.J., 1968, pp. 1987–1990.
26. Uni-Bell PVC Pipe Association: *Handbook of PVC Pipe Design and Construction*, Dallas, Tex., 1982.
27. Watkins, R. K.: "Design of Buried, Pressurized Flexible Pipe," ASCE National Transportation Engineering Meeting, Boston, Mass., July 1970, App. C.
28. Watkins, R. K., and A. P. Moser: "Response of Corrugated Steel Pipe to External Soil Pressures," Highway Research Record 373, 1971, pp. 88–112.
29. Watkins, R. K., A. P. Moser, and R. R. Bishop: "Structural Response of Buried PVC Pipe," *Modern Plastics*, 1973, pp. 88–90.
30. Watkins, R. K., and A. B. Smith: "Ring Deflection of Buried Pipe," *Journal AWWA*, vol. 59, no. 3, March 1967.
31. Watkins, R. K., and M. G. Spangler: "Some Characteristics of the Modulus of Passive Resistance of Soil—A Study in Similitude," *Highway Research Board Proceedings*, vol. 37, 1958, pp. 576–583.
32. White, H. C., and J. P. Layer: "The Corrugated Metal Conduit as a Compression Ring," *Highway Research Board Proceedings*, vol. 39, 1960, pp. 389–397.
33. Wilson, E.: "Solid SAP, A Static Analysis Program for Three Dimensional Solid Structures," SESM Report 71-19, Structural Engineering Laboratory, University of California, Berkeley, Ca., 1971.
34. Wong, K. S., and J. M. Duncan: "Hyperbolic Stress-Strain Parameters for Nonlinear Finite Element Analysis of Stresses and Movements in Soil Masses," Report no. TE-74-3, Office of Research Services, University of California, Berkeley, Ca., 1974.
35. Zienkiewitcz, O. C.: "The Finite Element Method," 3d ed., McGraw-Hill Book Company, New York, 1977.

4

Design of Pressure Pipes

The design methods for buried pressure pipe installations are somewhat similar to the design methods for gravity pipe installations which were discussed in Chap. 3. There are two major differences:

1. Design for internal pressure must be included.
2. Pressure pipes are normally buried with less soil cover so the soil loads are usually less.

Included in this chapter are specific design techniques for various pressure piping products. Methods for determining internal loads, external loads, and combined loads are given along with design bases.

Pipe Wall Stresses and Strains

The stresses and resulting strains arise from various loadings. For buried pipes under pressure, these loadings are usually placed in two broad categories: internal pressure and external loads. The internal pressure is made up of the hydrostatic pressure and the surge pressure. The external loads are usually considered to be those caused by external soil pressure and/or surface (live) loads. Loads due to differential settlement, longitudinal bending, and shear loadings are also considered to be external loadings. Temperature induced stresses may be considered to be caused by either internal or external effects.

Hydrostatic pressure

Lame's solution for stresses in a thick-walled circular cylinder is well known. For a circular cylinder loaded with internal pressure only, those stresses are as follows:

Tangential stress: $\sigma_t = \dfrac{P_i a^2 (b^2/r^2 + 1)}{b^2 - a^2}$

Radial stress: $\sigma_r = \dfrac{P_i a^2 (b^2/r^2 - 1)}{b^2 - a^2}$

where P_i = internal pressure
a = inside radius
b = outside radius
r = radius to point in question

The maximum stress is the tangential stress σ_t and occurs at $r = a$ (Fig. 4.1). Thus,

$$\sigma_{max} = (\sigma_t)_{r=a} = \frac{P_i a^2 (b^2/a^2 + 1)}{b^2 - a^2}$$

or

$$\sigma_{max} = \frac{P_i (b^2 + a^2)}{b^2 - a^2} \tag{4.1}$$

For cylinders (pipe) where $a \approx b$ and $b - a = t$,

$$(b^2 - a^2) = (b + a)(b - a) = \overline{D}t \tag{4.1a}$$

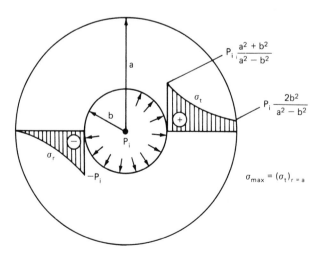

Figure 4.1 Thick-walled cylinder with internal pressure.

where $\overline{D}$ = average diameter = $(b + a)$
    t = thickness = $(b - a)$

Also, $(b + a)^2 = \overline{D}^2 = b^2 + a^2 + 2ab$. Thus,

$$b^2 + a^2 = \overline{D}^2 - 2ab \; \overline{D}^2 - 2r^2 = \overline{D}^2 - \frac{\overline{D}^2}{2} \qquad (4.1b)$$

Equation (4.1) can be rewritten using equations (4.1a) and (4.1b) as follows:

$$\sigma_{max} = \frac{P_i \, (\overline{D}^2/2)}{\overline{D}t} = \frac{P_i\overline{D}}{2t} \qquad (4.1)$$

Equation (4.2) is recognized as the equation for stress in a thin-walled cylinder (Fig. 4.2). This equation is sometimes called the Barlow formula, but is just a reduction from Lame's solution. This equation is the form most often recognized for calculating stresses due to internal pressure P_i.

If the outside diameter (D_o) is the reference dimension, Eq. (4.2) can be put into another form by introducing $\overline{D} = D_o - t$. That is, the average diameter is equal to the outside diameter minus thickness. Equation (4.2) becomes

$$\sigma_{max} = \frac{P_i \, (D_o - t)}{2t} \qquad (4.2)$$

Certain plastic pipe specifications refer to a dimension ratio (DR) or a standard dimension ratio (SDR), where

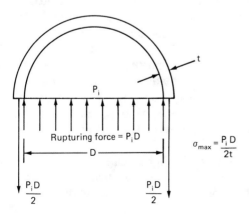

Figure 4.2 Free-body diagram of half section of pipe with internal pressure.

$$DR = \frac{D_o}{t} \quad \text{or} \quad SDR = \frac{D_o}{t}$$

Both DR and SDR are defined the same. However, SDR often refers to a preferred series of numbers that represented D_o/t for standard products. By introducing $D_o/t = SDR$ into equation (4.3), it can be rewritten as follows:

$$\sigma_{max} = \left(\frac{P_i}{2}\right)(SDR - 1) \tag{4.4}$$

The above equation may be expressed as

$$\frac{2\sigma_{max}}{P_i} = SDR - 1 \tag{4.5}$$

Equation (4.5) is often referred to as the ISO (International Standards Organization) equation for stress due to internal pressure. However, this basic equation has been known to engineers for more than a century and was originally given by Lamé in "Leçons sur la théorie de l'élasticité," Paris 1852. Obviously, ISO is a relative newcomer and should not be given credit for Lamé's work.

To calculate these tangential stresses in the pipe wall produced by internal pressure, either Eq. (4.2) or Eq. (4.4) are often suggested by the manufacturer or by national standards. All forms are derived from Lamé's solution and will produce comparable results.

Surge pressure

Pressure surges are often divided into two categories: transient surges and cyclic surges. Cyclic surging is a regularly occurring pressure fluctuation produced by action of such equipment as reciprocating pumps, undamped pressure control valves, oscillating demand, or other cyclic effects. Cyclic surges may cause fatigue damage and should be designed out of the system.

Transient surges are just that: transient in nature, occur over a relatively short time, and occur between one steady state and another. A transition surge may occur and the system then returns to the same steady state as before the surge. Transient surges are usually not cyclic in nature although they may be repetitive. A transient surge is often referred to as water hammer.

Any action in a piping system that results in a change in velocity of the system is a potential cause of a water-hammer surge. A partial listing of some typical causes of water hammer is given below.

1. Changes in valve settings (accidental or planned)

2. Starting or stopping of pumps

3. Unstable pump or turbine characteristics

The magnitude of water-hammer pressures generated by a given change in velocity depends on (1) the geometry of the system, (2) the magnitude of the change in velocity, and (3) the speed of the water-hammer wave for the particular system. These variables are expressed quantitatively as:

$$\Delta H = \left(\frac{a}{g}\right)\Delta V \tag{4.6}$$

where ΔH = surge pressure, feet of water
    a = velocity of pressure wave, ft/s
    g = acceleration due to gravity (32.17 ft/s^2)
    ΔV = change in velocity of fluid, ft/s

The pressure rise, in lb/in^2, may be determined by multiplying Eq. (4.6) by 0.43 lb/in^2 per feet of water as follows:

$$\Delta P = \left(\frac{a}{g}\right)\Delta V(0.43) \tag{4.7}$$

The wave speed is dependent upon:
1. Pipe properties
    a. Modulus of elasticity
    b. Diameter
    c. Thickness
2. Fluid properties
    a. Modulus of elasticity
    b. Density
    c. Amount of air, and so forth

These quantities may be expressed as

$$a = \frac{12\sqrt{K/\rho}}{\sqrt{1 + (K/E)(D/t)C_1}} \tag{4.8}$$

where a = pressure wave velocity, ft/s
    K = bulk modulus of water, lb/in^2
    ρ = density of water, slugs/ft^3
    D = internal diameter of pipe, in
    t = wall thickness of pipe, in
    E = modulus of elasticity of pipe material, lb/in^2
    C_1 = constant dependent upon pipe constraints ($C_1 = 1.0$ for pipe with expansion joints along its length)

For water at 60°F, Eq. (4.8) may be rewritten by substituting $\rho = 1.938$ slug/ft^3 and $K = 313{,}000$ lb/in^2.

$$a = \frac{4822}{\sqrt{1 + (K/E)(D/t)C_1}} \tag{4.9}$$

Equations (4.6), (4.7), and (4.8) can be used to determine the magnitude of surge pressure that may be generated in any pipeline. The validity of the equations has been shown through numerous experiments.

Figure 4.3 is a plot of the pressure rise in lb/in^2 as a function of velocity change for various values of wave speed. Tables 4.1 and 4.2 give the calculated wave speed according to Eq. (4.8) for ductile iron and PVC pipe, respectively. In general, wave speeds vary from 3000 to 5000 ft/s for ductile iron and from 1200 to 1500 for PVC pipes.

Example   Determine the magnitude of a water-hammer pressure wave induced in a 12-in class 52 ductile iron pipe and in a class 150 PVC pipe if the change in velocity were 2 ft/s.

solution   From Tables 4.1, 4.4, and Fig. 4.3:

Pipe	Wave speed, ft/s
Class 52 DI	4038
Class 150 PVC	1311

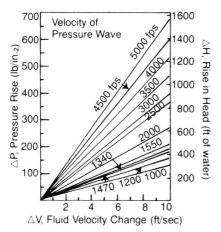

Figure 4.3   Water hammer surge calculation.

TABLE 4.1   Water-Hammer Wave Speed for Ductile Iron Pipe in ft/s*

ft/s

				Class			
Size	50	51	52	53	54	55	56
4	—	4409	4452	4488	4518	4544	4567
6	4206	4265	5315	4358	4394	4426	4454
8	4085	4148	4202	4248	4289	4324	4356
10	3996	4059	4114	4162	4205	4242	4276
12	3919	3982	4038	4087	4130	4169	4205
14	3859	3921	3976	4024	4069	4108	4144
16	3783	3846	3902	3952	3998	4039	4076
18	3716	3779	3853	3887	3933	4038	4014
20	3655	3718	3776	3827	3874	3917	3957
24	3550	3614	3671	3723	3771	3815	3855
30	3387	3472	3547	3615	3676	3731	3782
36	3311	3409	3495	3571	3638	3700	3755
42	3255	3362	3456	3539	3612	3678	3737
48	3207	3323	3424	3512	3590	3659	3721
54	3201	3320	3423	3512	3591	3599	3724

TABLE 4.2   Water Hammer Wave Speed for PVC Pipe in ft/s*

	(AWWA C900) Class			Pressure rated PVC SDR			
Size	100	150	200	21	26	32.5	41
4	1106	1311	1496	1210	1084	967	859
6	1106	1311	1496	1210	1084	967	859
8	1106	1311	1496	1210	1084	967	859
10	1106	1311	1496	1210	1084	967	859
12	1106	1311	1496	1210	1084	967	859

*AWWA C-150; Water at 60°F.

The resulting pressure surges are

Pipe	Surge pressure (lb/in^2)
Class 52 DI	105
Class 150 PVC	35

Some appropriate rules of thumb for determining maximum pressure surges are listed below in lb/in^2 surge per 1 ft/s change in velocity.

Pipe	Surge pressure rise lb/in^2 (1 ft/s velocity change)
Steel pipe	45
DI (AWWA C-150)	50
PVC (AWWA C-900)	20
PVC (pressure rated)	16

Since velocity changes are the cause of water-hammer surge, proper control of valving may eliminate or minimize water hammer. If fluid approaching a closing valve is able to sense that closing and adjust its flow path accordingly, then the maximum surge pressure as calculated from Eq. (4.6) may be avoided. In order to accomplish this, the flow must not be shut off any faster than it would take a pressure wave to be initiated at the beginning of valve closing and return again to the valve. This is called the "critical time" and is defined as the longest ellapsed time before final flow stoppage that will still permit this maximum pressure to occur. This is expressed mathematically as

$$T_{CR} = \frac{2L}{a}$$

where T_{CR} = critical time
     L = distance within the pipeline that the pressure wave moves before it is reflected back by a boundary condition, ft
     a = velocity of pressure wave for the particular pipeline, ft/s

Thus, the critical time for a line leading from a reservoir to a valve 3000 ft away for which the wave velocity is 1500 ft/s is

$$T_{CR} = \frac{2(3000)\text{ft}}{1500 \text{ ft/s}} = 4 \text{ s}$$

Unfortunately, most valve designs (including gate, cone, globe, and butterfly valves) do not cut off flow proportionate to the valve-stem travel (see Fig. 4.4).

This illustrates how the valve stem, in turning the last portion of its travel, cuts off the majority of the flow. It is extremely important, therefore, to base timing of valve closing on the "effective closing time" of the particular valve in question. This "effective time" may be taken as about one-half of the actual valve closing time. The effective time is the time that should be used in water-hammer calculations. Logan Kerr[9] has published charts that allow calculation of the percent of maximum surge pressure obtained for various valve closing characteristics.

There is one basic principle to keep in focus in design and operation of pipelines. Surges are related to changes in velocity. The change in pressure is directly related to the change in velocity. Avoiding sudden changes in velocity will generally avoid serious water-hammer surges. Taking proper precautions during initial filling and testing of a pipeline can eliminate a great number of surge problems.

In cases where it is necessary to cause sharp changes in flow velocity, the most economical solution may be a relief valve. This valve

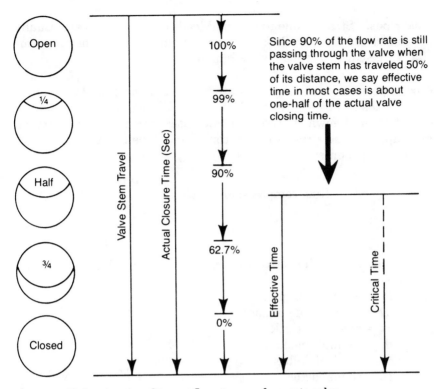

Figure 4.4 Valve stem travel versus flow stoppage for a gate valve.

opens at a certain preset pressure and discharges the fluid to relieve the surge. Such valves must be carefully designed and controlled to be effective.

Surge tanks can also be designed to effectively control both positive and negative surges. In general, they act as temporary storage for excess liquid that has been diverted from the main flow to prevent overpressures, or as supplies of fluid to be added in the case of negative pressures.

External loads

External earth loads and live loads induce stresses in pipe walls. Methods for calculating these loads were discussed in Chap. 2 and design procedures for external loads were discussed in Chap. 3. These loads and their effects should be considered in pressure-pipe installation design. Often stresses due to external loads are secondary in nature, but can be the primary controlling factor in design.

Rigid pipes. Stresses due to external loads on rigid pipes are usually not considered directly. Strength for rigid pipe is determined in terms of a "three-edge" test load (see Chap. 3).

Flexible pipes. Stresses in the wall of a flexible pipe produced by external loads can be easily calculated if the vertical load and resulting deflection are known. Methods for calculating the deflection are given in Chap. 3. These stresses can be considered to be made up of the following components:

$$\text{Ring compression stress:} \quad \sigma_c = \frac{P_v D}{2t} \qquad (4.10)$$

and

$$\text{Bending stresses:} \quad \sigma_b = D_f E\left(\frac{\Delta y}{D}\right)\left(\frac{t}{D}\right) \qquad (4.11)$$

where P_v = vertical soil pressure
   D = pipe outside diameter
   t = pipe wall thickness
   E = Young's modulus
   Δy = vertical deflection
   D_f = shape factor

This factor is a function of pipe stiffness as is indicated by Table 4.3. Generally, the lower the stiffness, the higher the D_f factor. Other parameters such as pipe zone soil stiffness and compaction techniques have an influence on this factor, but the values listed in Table 4.3 are recommended design values for proper installations.

TABLE 4.3   Pipe Stiffness

(F/y) (lb/in^2)	D_f
5	15
10	8
20	6
100	4
200	3.5

Total circumferential stress can be obtained by the use of the following:

$$\sigma_T = \sigma_p + \sigma_c + \sigma_b$$

where σ_T = total stress

σ_p = stress due to internal pressure (static and surge)

σ_c = ring compression stress

σ_b = stress due to ring deflection (bending)

This total stress may or may not be necessary to consider in design (see next section on combined loading).

Combined loading

A method of analysis which considers effects due to external loads and internal pressure acting simultaneously is called a "combined loading analysis."

Rigid pipes. For rigid pressure pipe such as cast iron or asbestos cement, the combined loading analysis is accomplished in terms of strength. The following procedure was originally investigated and suggested by Professor W. J. Schlick of Iowa State University. It has since been verified by others.

Schlick showed that, if bursting strength and the three-edge bearing strength of a pipe are known, the relationship between the internal pressures and external loads, which will cause failure, may be computed by means of the following equation:

$$w = W\sqrt{\frac{P - p}{P}} \qquad (4.12)$$

where w = three-edge bearing load at failure under combined internal and external loading, lb/ft

W = three-edge bearing strength of pipe with no internal pressure, lb/ft

P = burst strength of pipe with no external load, lb/in^2

p = internal pressure at failure under combined internal and external loading, lb/in^2

Schlick's research was carried out on cast-iron pipe and was later shown to apply to asbestos-cement pipe. Neither of these piping materials are currently available in the United States. However, these materials are available in other countries.

An example of the Schlick method of combined loading design for a rigid pipe is as follows: Suppose a 24-in asbestos-cement water pipe has a three-edge bearing strength of 9000 lb/ft and a bursting strength of 500 lb/in^2. Figure 4.5 shows graphs of Eq. (4.12) for various strengths of asbestos-cement pipes. The curve for this particular pipe is labeled 50. If this pipe were subjected in service to a 200 lb/in^2 pressure (including an allowance for surge) times a safety factor, this

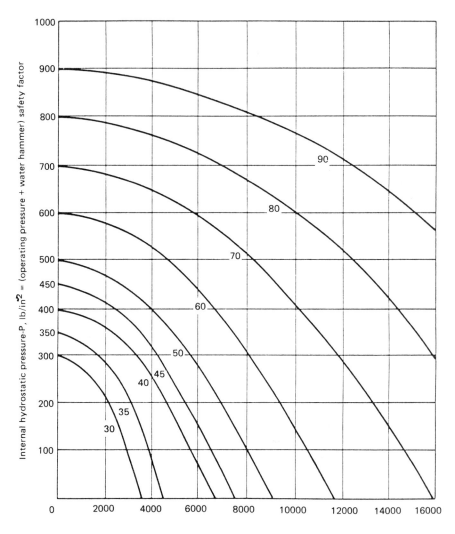

External crush load (vee-shaped bearing) W, lb/ft

$$W_T = (\text{earth load} + \text{live load}) \ \frac{\text{Bedding factor}}{\text{Safety factor}}$$

Figure 4.5 Combined loading curves for 24-in asbestos-cement pressure pipe. (*Reprinted from ANSI/AWWA C403-84,[3] by permission. Copyright 1984, American Water Works Association.*)

graph shows the pipe would have a three-edge bearing strength, in service, of 7000 lb/ft; for an internal service pressure of 400 lb/in² the three-edge bearing strength would be 4000 lb/ft; and so on. The three-edge bearing strength must be multiplied by an appropriate load factor to obtain the resulting supporting strength of the pipe when actually installed.

Flexible pipes. For most flexible pipes, such as steel, ductile iron, and thermal plastic, a combined loading analysis is not necessary. For these materials, the pipe is designed as if external loading and internal pressure were acting independently. Usually, pressure design is the controlling factor. That is, a pipe thickness or strength is chosen on the basis of internal pressure and then an engineering analysis is made to make sure the chosen pipe will withstand the external loads.

An exception to the above statement is fiberglass reinforced thermal-setting resin plastic (FRP) pipe. This particular type of pipe is designed on the basis of strain. The total combined strain must be controlled to prevent environmental stress cracking. A recommended design procedure is given in Appendix A of AWWA C-950. The total combined strain in this case is the bending strain plus the strain due to internal pressure. Some FRP pipe manufacturers recommend all components of strain be added together to get the total maximum strain. The following is a list of some loadings or deformations that produce strain.

1. Internal pressure
2. Ring deflection
3. Longitudinal bending
4. Thermal expansion/contraction
5. Shear loadings
6. Poisson's effects

Longitudinal stresses

Pressure pipe as well as gravity-flow pipes are subject to various soil loadings and nonuniform bedding conditions that result in longitudinal bending or beam action. This subject was discussed in Chap. 2. Pressure pipes may also have longitudinal stresses induced by pressure and temperature which should be given proper consideration by the engineer responsible for installation design.

Poisson's effect. Engineers who deal with mechanics of materials know that applied stresses in one direction produce stress and/or strains in a perpendicular direction. This is sometimes called the Poisson effect. A pipe with internal pressure p has a circumferential stress σ_p. The associated longitudinal stress σ_ν is given by the following equation:

$$\sigma_\nu = \nu\sigma_p$$

where σ_ν = longitudinal stress
   ν = Poisson's ratio for pipe material
   σ_p = circumferential stress

The above equation is based on the assumption that the pipe is restrained longitudinally. This assumption is valid for pipes with rigid joints or for pipes with extra long lengths even if joined with slip joints such as rubber ring joints. Studies have shown that soil-pipe friction can cause complete restraint in approximately 100 ft. For shorter lengths with slip joints, since the restraint will not be complete, the longitudinal stress will be less than predicted by the above equation. For reference, some values of Poisson's ratio ν and Young's modulus E are listed in Table 4.4.

TABLE 4.4

Material	Modulus	Poisson's ratio
Steel	30×10^6 lb/in²	0.30
Ductile iron	24×10^6 lb/in²	0.28
Copper	16×10^6 lb/in²	0.30
Aluminum	10.5×10^6 lb/in²	0.33
PVC	4×10^6 lb/in²	0.45
Asbestos-cement	3.4×10^6 lb/in²	0.30
Concrete	$57,000\ (f_c')^{1/2}$ lb/in²*	0.30

*Where f_c' = 28-day compressive strength.

Temperature effects. Expansion or contraction due to temperature increase or decrease can induce longitudinal stress in the pipe wall. As with the Poisson stresses discussed above, these stresses are based on longitudinal restraint. The longitudinal stress due to temperature is given by the equation

$$\sigma_T = -\alpha(\Delta T)E$$

where σ_T = longitudinal stress due to temperature
   α = linear coefficient of expansion
   ΔT = temperature change
   E = Young's modulus for pipe material

An example of a situation that would cause such a stress follows: Consider a welded steel line which is installed and welded during hot summer days and later carries water at 35°F. The resulting ΔT will be substantial as will the resulting stress. Additional information on temperature induced stresses in welded steel pipe can be found in AWWA M-11 Steel Pipe Manual and in other AWWA standards on welded steel pipe.

Pipe thrust. Longitudinal stresses due to pipe thrust will be present when a piping system is self-restraining with welded, cemented, or locked-joint joining systems. For example, at a valve, when the valve is closed, the thrust force is equal to pressure P times area A. The same force is present at a 90 degree bend.

$$\text{Thrust} = \text{pressure} \times \text{area}$$

$$PA = P\pi r^2$$

The stress due to this thrust is given by the following:

$$\sigma_{th} = PA/2\pi rt = P\pi r^2/2\pi rt = Pr/2t$$

where σ_{th} = longitudinal stress due to thrust
  T = thrust force = $P\pi r^2$
  P = internal pressure plus surge pressure
  r = average radius of pipe
  t = thickness of pipe wall

Stress risers. The pipe system designer should always be aware of stress risers which will amplify the stresses. Stress risers occur around imperfections such as cracks, notches, and ring grooves. They are also present near changes in diameters such as the bell area. Designs that overlook stress risers can and have led to piping system failure. In a welded bell and spigot-type joint, the longitudinal tensile stresses are not passed across the joint without inducing high bending moments and resulting bending stresses. These bending stresses have been shown to be as high as seven times the total longitudinal stress in a straight section. For this example, the maximum longitudinal stress is given by the following:

$$(\sigma_L)_{max} = \beta(\sigma_{Lb} + \sigma_{Lv} + \sigma_{LT} + \sigma_{th})$$

where $(\sigma_L)_{max}$ = maximum longitudinal stress
  β = stress riser
  σ_{Lb} = stress due to longitudinal beam action
  σ_{Lv} = longitudinal stresses due to Poisson's effect

σ_{LT} = longitudinal stresses due to temperature
σ_{th} = longitudinal stresses due to thrust

Design Bases

Each piping material has criteria for design such as a limiting stress and/or a limiting strain. Also, each product may be limited as to specific application in terms of fluids it may carry or in terms of temperature. Usually these limiting conditions are translated into codes, standards, and specifications. Such specifications will deal with specific acceptable applications, permissible soil load, or depth of cover, internal pressure, safety factors, methods of installation, life, and in some cases, ring deflection. The limiting parameters for a given product when considered together form the basis for design.

Rigid pipes

The use of pressure pipe constructed wholly from rigid material is rapidly becoming history. Cast-iron pipe has been replaced with ductile iron, which is considered to be flexible. Asbestos-cement pressure pipe is still in production, but is rapidly losing out in the marketplace. Concrete pressure pipe, which is really steel pipe with a concrete liner and a concrete or cement grout coating, is usually considered to be rigid.

Asbestos cement. Design information for asbestos-cement pressure pipe can be found in AWWA C-401 and in AWWA C-403. A combined load analysis using the Schlick formula is required. This method is discussed under the combined loading section of this chapter. Eq. (4.12) is repeated here.

$$w = W\sqrt{\frac{P - p}{P}}$$

(4.12)

or

$$p = P\left[1 - \left(\frac{w}{W}\right)^2\right]$$

(4.13)

It is generally considered desirable to use the thick-walled formula for ratios of diameter-to-thickness exceeding 10. Equations (4.1) and (4.2) are the thick-wall formula and the thin-wall formula for hoop stress σ_t. The parameters W and P are determined experimentally. With these values, one can determine combinations of internal pressures p and external crush loads w that are necessary to cause failure. In addition, the design pressure will require an appropriate safety factor.

Normally if surges are present, the maximum design (operating) pressure is one-fourth of the pressure to cause failure. If surges are not present, the operating design pressure is four-tenths the failure pressure.

The design crush load is equal to the expected earth load plus live load times the safety factor (usually 2.5) and divided by a bedding factor (see Chap. 3 for bedding factors).

W = (earth load + live load)(safety factor)/bedding factor

Design curves are given in AWWA C-401 and AWWA C-403 (see Fig. 4.5 for an example). The designer enters the graph by locating the appropriate design pressure on the vertical axis and the appropriate external crush load on the horizontal axis. The intersection of these grid lines locate the appropriate pipe curve. If the intersection is between curves, choose the next higher curve and the associated strength pipe.

Reinforced concrete.   Reinforced concrete pressure pipe is of four basic types:

1. Reinforced steel cylinder (AWWA C-300-74)

2. Prestressed steel cylinder (AWWA C-301-72)

3. Reinforced noncylinder (AWWA C-302-74)

4. Pretensioned steel cylinder (AWWA C-303-70)

For rigid pipes discussed up to this point, the performance limits have been described in terms of rupture of the pipe wall due to either internal or external loads or some combination thereof being greater than the strength of the pipe. Performance limits for reinforced concrete pipe are described in terms of design conditions such as zero compression stress, and so forth. Generally, the design of reinforced concrete pressure pipe requires the consideration of two design cases:

1. A combination of working pressure and transient pressure and external loads

2. A combination of working pressure and external load (earth plus live load)

Reinforced steel cylinder pipe is designed on a maximum combined stress basis. The procedure is to calculate stresses in the steel cylinder and steel reinforcement produced by both the external loads and internal pressure. The combined stress at the crown and invert must be equal to or less than an allowable tensile stress for the reinforcing steel and steel cylinder. See Appendix A of AWWA C-300 for details concerning design procedure.

Prestressed concrete pipe is designed for combinations of internal and external loads by the following cubic equation:

$$w = w_o \sqrt[3]{\frac{P_o - p}{P_o}} \tag{4.14}$$

where P_o = internal pressure which overcomes all compression in concrete core, when no external load is acting, lb/in^2

W_o = 90 percent of three-edge bearing load which causes incipient cracking in core when no internal pressure is acting, lb/ft

p = maximum design pressure in combination with external loads (not to exceed $0.8P_o$ for lined cylinder pipe)

w = maximum external load in combination with design pressure

The value of W_o can be determined by test and the value of P_o can be either determined by test or calculated. With these parameters known, w and p can be calculated using Eq. (4.14) in a manner that is similar to the use of the Schlick formula for asbestos-cement pipe. Further information concerning the combined loading analysis using the cubic parabola Eq. (4.14) is available in AWWA C-301.

Pretensioned concrete cylinder pipe is considered by many to be a rigid pipe. Truly, it does not meet the definition of a flexible pipe (must be able to deflect 2 percent without structural distress). The limiting design deflection for pretensioned concrete cylinder pipe ranges from 0.25 percent to 1.0 percent. AWWA Manual M9 indicates that this type of pipe is semirigid. However, the recommended design procedure found in AWWA C-303 is based on flexible pipe criteria. The recommended procedure is to limit stresses in steel reinforcement and the steel cylinder to 16,500 lb/in^2 or 50 percent of the minimum yield, whichever is less. The stiffness of the pipe must be sufficient to limit the ring deflection to not more than $D^2/4000$. D is the nominal inside diameter of the pipe in inches.

Flexible pipes

Thermoplastic. All plastics are, at some stage, soft and pliable and can be shaped into desired forms usually by the application of heat, pressure, or both. Some can be cast. Thermoplastics soften repeatedly when heated and harden when cooled. At high enough temperatures, they may melt; and at low enough temperatures, they may become brittle. A few familiar examples of thermoplastics used for pipe are polyvinyl chloride (PVC), polyethylene (PE), acrylonitrile butadiene styrene (ABS) polybutylene (PB), and styrene rubber (SR). No matter what the type of thermoplastic pressure pipe, there is common terminology. A detailed review of some will be made, but the design engineer should become familiar with these terms as they are somewhat unique to the plastic pressure-pipe industry.

Plastic pressure-pipe terminology

Stress regression

Cell classification

Quick-burst strength

Hydrostatic-design basis

Hydrostatic-design stress

Service factor

Safety factor

Pressure rating

Pressure class

SDR

DR

PVC compounds. The original method for classifying PVC compounds was by types and grades, for example for PVC:

1. Type I, grade 1: Normal impact, very high chemical resistance and highest requirements for mechanical material strength.

   Type I, grade 1 compounds are by far the predominant material used today for pipe. Other types and grades of compounds are as follows:

2. Type I, grade 2: Essentially the same properties as grade 1 but possesses lower requirements for chemical resistance. Grade 1 has about 5 percent higher hoop stress based on 50-year strength.

3. Type II, grade 1: High impact strength but sacrifice chemical resistance and tensile strength.

4. Type III, grade 1: Medium impact strength, low chemical resistance.

While this terminology still persists, the current definition of PVC compounds is defined in the most current edition of ASTM D1784, the standard specification for "rigid poly(vinyl chloride) (PVC) compounds and chlorinated poly(vinyl chloride) (CPVC) compounds." This specification defines the physical characteristics of the compound with a five digit cell-class numbering system and a letter suffix describing chemical resistance.

The old type and grade compound system is now expressed in cell classification as follows:

Type I, grade 1: 12454B

Type I, grade 2: 12454C

Type II, grade 1: 14333D

Type III, grade 1: 13233

Type IV, grade 1: 23447B

The following is a brief review of what this numbering matrix plus a letter, that is, 12454B defines.

First number: Material identification (PVC homo polymer)

Second: Impact strength (izod minimum) (0.65 ft lb/in)

Third: Tensile strength (7000 lb/in^2 minimum)

Fourth: Modulus of elasticity (in tension 400,000 lb/in^2 minimum)

Fifth: Deflection temperature under load (158°F minimum)

Letter: Chemical resistance as defined in Table 2 of ASTM D1784.

As indicated, the PVC compound most commonly used for water (pressure) pipe application is:

Old designation: type I, grade 1

Current designation: 12454 B

In late 1980, ASTM approved yet another standard for identifying PVC compounds. ASTM D3915 utilizes a similar cell-class system as ASTM D1784, but has deleted the letter suffix and substituted a hydrostatic-design basis cell. To date, this system has not been adopted in any PVC pipe standards.

Hydrostatic design basis.   ASTM D2837 establishes the "standard method for obtaining hydrostatic-design basis for thermoplastic pipe materials."

The procedure for establishing a hydrostatic-design basis is as follows:

1. Classify PVC pipe compound per ASTM D1784 cell classification, that is, 12454B

2. Conduct long-term static pressure tests on pipe

3. Submit data to the hydrostatic-design committee of PPI for analysis

4. Determine the long-term hydrostatic strength (LTHS) [LTHS is the extrapolated hoop stress that will produce failure in 100,000 hr. (11.4 yr)]

5. Determine the hydrostatic-design basis (HDB) by categorizing the LTHS per ASTM D2837, which also involves projections to 50 yr

The pressure test data, when presented on a log-log plot, form a straight line. It is called a "stress regression curve" (see Fig. 4.6 for stress regression curve for PVC). This "declining" curve does not represent a loss of strength with time. It does show that the higher the stress, the shorter the life, conversely the lower the stress, the longer the life. The line relates the life of the pipe to the level of stress in the pipe wall due to internal pressure. It is a series of test data points. For example, for a given stress or pressure, failure will occur in a given time. To establish the regression line, tests must be conducted such that individual failures occur from 10 to 10,000 hr (1.14 yr). The line is for static pressure only and temperature controlled at 73.4°F.

For PVC pipe, long-term static pressure tests have been carried out

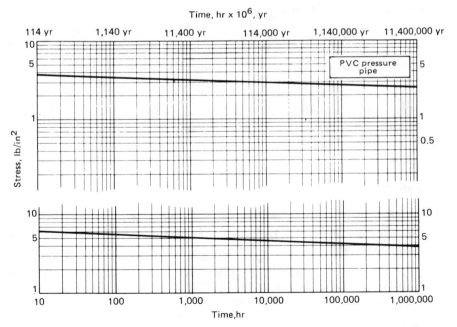

Figure 4.6 Stress regression line. (*Reprinted by permission of the Plastic Piping Institute.*)

over more than 200,000 hr (22.8 yr) that confirm the validity of establishing long-term hydrostatic strength on the basis of log-log straight-line extrapolations.

Hydrostatic design stress. The hydrostatic-design stress (HDS) is defined in ASTM D2241 as follows: "The estimated maximum tensile stress in the wall of the pipe in the circumferential orientation due to the internal hydrostatic water pressure that can be applied continuously with a high degree of certainty that failure will not occur."

The ASTM specifications for PVC, PE, and ABS pipe indicate the hydrostatic-design basis and hydrostatic-design stress for these materials. A comparison of one type and grade designation of each material reveals the following:

	HDB	HDS
PVC 1120*	4000	2000
PE 3406	1260	630
ABS 1316	3200	1600

*Equivalent to PVC cell classification 12454B per ASTM D-1784.

The higher HDB and HDS for PVC 1120 partially explains its wide acceptance for plastic water pressure pipe. A complete listing for these values for PVC, PE, and ABS is as follows in Table 4.5.

Pressure rated pipe. For the purpose of reviewing the plastic pressure-pipe design procedure, PVC pipe standard ASTM D2241 "Standard Specification for Poly (Vinyl Chloride) (PVC) Plastic Pipe (SDR-PR)" will be considered. A similar procedure exists for other thermoplastic materials.

Throughout existing PVC standards and specifications for PVC pipe one still finds the older "type and grade" designation. For example, the most common designation for pressure pipe is PVC 1120. It can be defined as follows:

PVC: Polyvinyl chloride

First number: (1) Represents type of compound, in this case, type I.

Second number: (1) represents the compound grade, in this instance, grade 1.

Third and fourth numbers: (20) represent, the hydrostatic-design stress, in this case 2000 lb/in^2 divided by 100, and decimals that result are dropped.

TABLE 4.5

Material type	Hydrostatic-design basis (lb/in^2)	Hydrostatic-design stress (lb/in^2)
PVC ASTM D2241		
PVC 1120	4000	2000
PVC 1220	4000	2000
PVC 2120	4000	2000
PVC 4120	4000	2000
PVC 2216	3200	1600
PVC 2112	2500	1250
PVC 2110	2000	1000
Polyethylene ASTM D2239		
PE 2306	1260	630
PE 3306	1260	630
PE 3406	1260	630
PE 2305	1000	500
PE 1404	800	400
ABS ASTM D1527		
ABS 1208	1600	800
ABS 1210	2000	1000
ABS 1316	3200	1600
ABS 2112	2500	1250

The design basis for PVC pressure pipe meeting ASTM D2241 is a balance of forces (Fig. 4.7). The pressure P times the mean diameter $D - t$ equals the stress σ times twice the wall thickness t or it can be expressed as follows:

$$P(D - t) = \sigma \times 2t$$

or

$$\sigma = \frac{P(D - t)}{2t} \qquad (4.15)$$

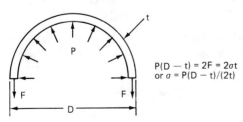

$$P(D - t) = 2F = 2\sigma t$$
$$\text{or } \sigma = P(D - t)/(2t)$$

Figure 4.7   Stress due to internal pressure.

where P = internal pressure, lb/in^2
σ = tensile strength, lb/in^2

The hydrostatic-design stress, HDS or σ in the equation, is the hydrostatic-design basis (HDB) times a service factor. HDB for PVC is 4000 lb/in^2 for water-pipe compounds. The service factor is defined in the appendix of ASTM D2241 and recommended by the Plastic Pipe Institute as equal to 0.5. (The inverse of the service factor is the safety factor, in this case 2). Thus, the long-term hydrostatic-design stress for PVC 1120 pressure-rated pipe meeting ASTM D2241 is 2000 lb/in^2 (HDB $\times$ 0.5 or HDB/2).

Equation (4.15) can be rearranged algebraically to reveal the term SDR (standard dimension ratio) or

$$\frac{D}{t} = \frac{\text{Average outside diameter}}{\text{Minimum wall thickness}}$$

This term is widely used in the thermoplastic pipe industry. Equation (4.15) can therefore be rearranged as follows:

$$2\sigma = \frac{P(D - t)}{t} \qquad \text{but} \qquad \frac{D}{t} = \text{SDR}$$

Therefore,

$$2\sigma = P(\text{SDR} - 1)$$

The term standard dimension ratio is referencing a preferred series of numbers. Also note that the pressure rating of a given SDR is the same no matter what the size, that is, 2-in and 12-in SDR 26 have the same pressure rating.

In review, the four basic ideas that are important to the designer of thermoplastic pressure pipe are:

1. The hydrostatic-design basis (HDB) for a given PVC pipe extrusion compound is established through long-term hydrostatic pressure testing for pipe extruded from that compound.

2. The hydrostatic design stress (HDS or S) is the stress in the pipe wall at which plastic pipe will perform indefinitely.

3. The service factor (0.5) times (or the safety factor, 2 to 1, divided into) the hydrostatic-design basis, equals the hydrostatic-design stress.

4. Plastic pipe does not lose strength with time.

AWWA standards. The first AWWA standard approved for plastic pipe was AWWA C-900 in 1975. AWWA C-900 is the AWWA standard for polyvinyl chloride (PVC) pressure pipe, 4 in through 12 in for water. This standard contains three pressure classes.

Class 100, DR 25

Class 150, DR 18

Class 200, DR 14

The term DR means the same as the standard dimension ratio, i.e.,

$$DR = \frac{\text{outside diameter}}{\text{minimum wall thickness}} = \frac{D}{t}$$

However, the values do not fall in the referenced ASTM preferred series. The design basis for AWWA C-900 differs from ASTM in two areas:

1. It has a higher safety factor.
2. It includes a surge allowance.

The design-basis equation in C-900 can be expressed in the following way:

$$2.5(PC + PS) = \frac{2t}{(D - t)}(HDB) \tag{4.16}$$

where 2.5 = safety factor
  PC = pressure class: 100 lb/in^2, 150 lb/in^2, 200 lb/in^2
  PS = surge allowance (lb/in^2) for instantaneous stoppage of flow of 2 ft/s
  t = minimum wall thickness, in
  D = outside diameter, in
  HDB = hydrostatic-design basis = 4000 lb/in^2

The actual surge allowances in AWWA C-900 PVC pipe that result from stoppage of flow of 2 ft/s are as follows:

Class 100, DR 25: 30 lb/in^2

Class 150, DR 18: 35 lb/in^2

Class 200, DR 14: 40 lb/in^2

Another design parameter included in AWWA and not in ASTM is the effect of sustained elevated temperatures on pressure and/or design stress. For sustained temperature of the pipe wall above 73°F the design stress should be reduced. This reduction is not necessary for short-term excursions of elevated temperatures but is for continuous service at a higher temperature. The recommended percentages of allowable pressure class for various elevated temperatures are as shown in the following table:

Maximum continuous service temperature, °F	Percentage of allowable pressure-class or design stress at 73°F
73	100
80	88
90	75
100	62
110	50
120	40
130	30
140	22

A review of AWWA C-900 and AWWA C-905 for PVC water pipe indicates approval of the following:

Compounds: 12544A or 12544B (formerly 1120)

Size range: 4 in through 12 in (AWWA C-900)

14 in through 36 in (AWWA C-905)

Pressure classes: 100, 150, and 200 lb/in² (AWWA C-900)

100, 125, 160, 165, 200, and 235 lb/in² (AWWA C-905)

DR: 25, 18, and 14 (AWWA C-900)

41, 32.5, 26, 21, and 18 (AWWA C-905)

In 1980, AWWA published a new manual, AWWA no. M23. This manual was a follow-up on the standard AWWA C-900 and covers the design and installation of PVC pipe meeting this standard. This manual provides a very thorough presentation on the design and installation of PVC pipes.

AWWA C-901 is the AWWA standard for polyethylene (PE) pressure pipe, tubing, and fittings, ½ in through 3 in for water. This standard is primarily for PE water service piping. AWWA C-901 can be summarized as follows:

Compounds: 2306, 2406, 3406, 3408

Size range: ½ in through 3 in

Diameters: ID base

OD base

Tubing

Pressure classes: 80, 100, 125, 150, 160, and 200 lb/in²

DR: 17.0, 15.0, 13.5, 11.5, 11.0, 9.3, 9.0, 7.3, 7.0, and 5.3

Note that AWWA C-901 provides for a wide selection of compounds and pressure classes. Safety factor is 2 and there is no routine test for each piece of pipe.

AWWA C-902 is the AWWA standard for polybutylene (PB) pressure pipe, tubing, and fittings, ½ in through 3 in for water. This standard is also primarily intended for service water piping. AWWA C-902 can be summarized as follows:

Compounds: Type II, grade 1, class B,

Type II, grade 1, class C

HDB = 2000 lb/in²; PB 2110

Size range: ½ in through 3 in

Diameters: ID base; OD base; and tubing

Pressure classes: 125, 160, and 200 lb/in²

DR: 17.0, 15.0, 13.5, 11.5, 11.0, and 9.0

Again, a wide selection is available. The safety factor is also 2 and no routine test is required for each piece of pipe.

Cyclical pressure surges.   In a water distribution system, surge conditions normally occur on a rather infrequent basis. However, if a system is operating out of control, frequent or cyclical pressure surges can occur. It is this type of condition that may require additional design considerations for plastic pipe. Research work has shown the following:

1. Plastic pipe possesses two life funds—static and dynamic (or hydrostatic and cyclic)

2. These funds are separate and independent of each other

3. The cyclic pressure life fund may be a critical parameter if the magnitude and frequency of surges are high

The following table (Table 4.6) shows some cyclical pressure research work done on 6-in class 150 PVC pipe:

This table shows that a system must be operating completely out of

TABLE 4.6 6-in Class 150 PVC Pipe

Sample	Outside diameter, in	Min. wall, in	Max. pressure, lb/in²	Max. stress, lb/in²	Cycles at failure
1	6.910	0.448	660	4760	7,163
2	6.909	0.444	480	3495	40,798
3	6.909	0.443	760	5546	2,851
4	6.908	0.448	650	4686	2,851
5	6.913	0.442	485	3550	27,383
6	6.911	0.436	600	4455	10,105
7	6.910	0.440	340	2500	78,403
8	6.904	0.452	340	2427	121,768
9	6.910	0.452	340	2429	91,475
10	6.910	0.443	200	1460	3,018,907
11	6.910	0.443	235	1715	983,200
12	6.910	0.444	440	3204	35,633
13	6.908	0.443	340	2481	119,971
14	6.910	0.447	200	1446	3,647,182
15	6.910	0.449	200	1439	2,563,538
16	6.910	0.442	400	2927	63,616
17	6.907	0.455	600	4254	3,587
18	6.906	0.450	680	4878	2,936

control to cause a problem. Class 150 PVC is designed for a total continuous pressure of 185 lb/in², i.e., 150 lb/in² + 35 lb/in². For a problem to occur at this design pressure, it would have to cycle more than a million times. The large magnitude of surges necessary to cause failure within a relatively few cycles is not a normal occurrence.

This does not mean to say that for PVC pressure pipe, or pressure pipe made of any other material, the control of surges in a water distribution is not imperative.

If the cyclical data for PVC pipe were plotted on a log plot, it would be a straight line (see Fig. 4.8).

$$C = (5.05 \times 10^{21}) \times (S^{-4.906}) \qquad (4.17)$$

where S = peak hoop stress, lb/in²
C = average number of cycles (surges) to failure

To select the appropriate PVC pipe for a new installation, the following steps can be taken:

1. Determine the years of service required.

2. Determine the peak surge pressure that will be allowed.

3. Determine the number of surges that will be allowed.

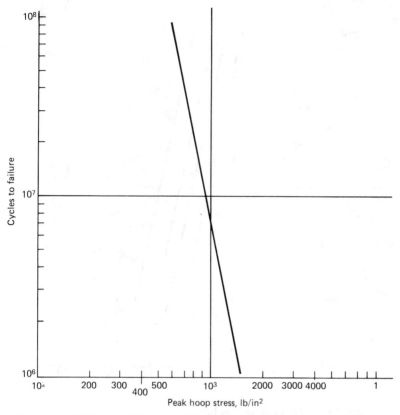

Figure 4.8 PVC pipe cyclic fatigue life large pressure cycles.

Based on these data, the number of surges expected during the system lifetime can be calculated (see example design curves, Fig. 4.9).

One can then determine the peak cyclic hoop stress allowed by modifying the previous equation as follows:

$$S' = \left(\frac{5.05 \times 10^{21}}{C'} \right)^{0.204} \tag{4.18}$$

where S' = required design-peak hoop stress, lb/in^2
     C' = anticipated number of cycles in life system

One may then determine the class of pipe required by determining the dimension ratio, DR, or (OD/t), using the following equation:

$$DR = \frac{2S'}{P} + 1 \tag{4.19}$$

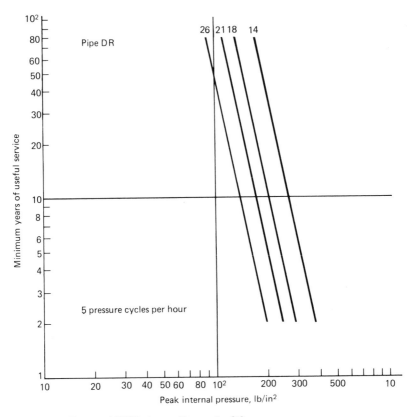

Figure 4.9 Expected PVC pipe cyclic service life.

where P = peak internal pressure (hydrostatic + surge), lb/in^2
  S' = required design peak hoop stress, lb/in^2
  DR = dimension ratio

These surge design equations represent useful tools in the analysis of fatigue failure due to cyclic surging; however, the engineer must appreciate the following limitations in use of these equations:

- The equations were developed empirically through testing of PVC pipe specimens only with large surges (at least 50 percent above base pressure or 25 percent above and below base pressure).

- The equations were developed for cycle frequencies in the range of 6 to 10 per minute. These equations provide no allowance for the stress relaxation phenomenon which may be experienced by PVC pipe at other cyclic frequencies. Based on other research efforts, it can be assumed that this design approach is conservative. Independent laboratory evaluation has demonstrated that AWWA C-900 class 150 PVC pipe can sustain in excess of 150 percent of the number of surge cycles predicted by these equations.

There are certain design applications, such as sewer forcemains, that may require the consideration of cyclic pressure life in any pipe product. A plot of cycles to failure for various peak pressures and PVC pipe-dimension ratios is provided in the second edition of the Uni-Bell *Handbook of PVC Pipe.*

Ductile iron. The design approach for ductile-iron pressure pipe is given in AWWA C-150. Various thickness classes are available. The required thickness is determined by considering stress due to internal pressure, ring deflection, and stresses due to earth loads separately and independently.

Calculations are made for the thicknesses required to resist the bending stress and the deflection due to trench load. The larger of the two is selected as the thickness required to resist trench load. Calculations are then made for the thickness required to resist the hoop stress of internal pressure. The larger of these is selected as the net design thickness. To this net thickness is added a service allowance and a casting tolerance to obtain the total calculated thickness. The standard thickness and the thickness class for specifying and ordering are selected from a table of standard class thicknesses. The reverse of the above procedure is used to determine the rated working pressure and maximum depth of cover for pipe of given thickness class.

Trench load P_v. Trench load is expressed as vertical pressure, lb/in², and is equal to the sum of earth load P_e and truck load, P_t.

Earth load P_e. Earth load is computed as the weight of the unit prism of soil with a height equal to the distance from the top of the pipe to the ground surface (prism load). The unit weight of backfill soil is taken to be 120 lb/ft³. If the designer anticipates additional loads due to frost, the design load should be increased accordingly.

Truck load P_t. The truck loads are computed using the surface load factors for a single AASHTO H-20 truck on unpaved road or flexible pavement, 16,000-lb wheel load and 1.5 impact factor.

The stress of the pipe invert produced by the total external loading is limited to 48,000 lb/in². This stress is the sum of the bending stress and the wall thrust stress. Wall stresses due to external loads are a function of this type of installation. Various installation types are considered in AWWA C-150 with associated tables for stresses as a function of depth of cover.

The ring deflection is limited to 3 percent. This is a design condition independent of wall stress. Most ductile-iron water pipes have a cement mortar lining. The three percent limitation is to protect that lining from cracking or spalling.

The wall stress due to internal pressure must be equal to or less than 21,000 lb/in^2. The yield stress in tension for ductile iron is approximately 42,000 lb/in^2. Thus a design stress of 21,000 is based on a safety factor of 2.0.

The above procedure is not based on a combined loading analysis and a combined loading criterion is not recommended. Each performance criteria is evaluated separately and the controlling parameter dictates the design thickness.

The following is a summary of the design bases for ductile iron pipe:

Soil loading

Prism load + truck load (H-20 × impact)

Deflection

(A) Iowa formula (3 percent limit)

Stress due to soil loading

(B) Calculate stress at invert (limit 48,000 lb/in^2)

Stress due to internal pressure

(C) $$S = \frac{PD}{2T} \text{ (limit 21,000 } lb/in^2)$$

P = working pressure + surge allowance

Design procedure

1. Calculate required thicknesses in steps (A), (B) and (C) above

2. Subtract service allowance (corrosion) of 0.08 in from thickness found in (A) and compare with thickness found in (B) and (C) (largest value controls)

3. Add service allowance (0.08 in) for minimum thickness

4. Add casting allowance for total thickness

The net effect of the above procedure is to include a service allowance for stress considerations but not for deflection. The rationale for not including it for deflection is that the deflection limit is not based on the ductile iron, but on the lining.

Steel pipe.   There are many types of steel pipe and many applications. AWWA M-11 (steel pipe design and installation) adequately covers the design and installation procedures. A brief review is included here.

Steel pipe is, for the most part, designed for internal pressure and installed in such a manner that other design considerations or limits are met. The basic design procedure is as follows:

Design for pressure (A)
Calculate stiffness (B)
Design soil system based on:
  1. Pipe stiffness

  2. Depth of cover (C)

  3. Performance limits

Pressure design.   Pressure design is based on the thin-wall pressure formula as follows:

$$S = \frac{pD}{2t} \quad \text{or} \quad t = \frac{pD}{2s}$$

where s = safe working stress (usually 50 percent of yield)
  p = working pressure plus calculated surge
  t = steel thickness

Determine stiffness.   If the pipe is not cement coated or lined, the stiffness is easily calculated.

$$\text{Stiffness} = EI = E\left(\frac{t^3}{12}\right)$$

where t = thickness determined from pressure design
  E = usually 30×10^6 lb/in^2

For cement-lined and/or coated-steel pipe, the stiffness will be available from the manufacturer or can be determined experimentally. For pipes that are lined after installation, only the steel should be considered in any stiffness calculation.

Soil system design.   The known parameter at this point in the design will be:

1. Pipe performance limits, usually two percent deflection

2. Depth of cover

3. Pipe stiffness

The parameters to be determined are pipe-zone soil type and soil density in pipe zone—embedment techniques, and so forth.

Recommended procedures. Loads may be calculated by Marston's Iowa formula for flexible pipe or by the prism-load method. See Chaps. 2 and 3 for details.

Deflection is determined using Spangler's Iowa formula, Watkin's soil strain method or by the use of empirical data.

Manufacturers recommendations should be given serious consideration in this regard as many have developed tables for deflection from actual test data. Also, other standards, such as AWWA C-200 and AWWA C-206, provide useful and pertinent information regarding installation design.

Buckling. Many steel pipelines are extremely flexible and may be subject to buckling or collapse from external pressure or internal vacuum. The engineer should consider buckling in the design and take appropriate action to eliminate it. Vacuum relief values may be necessary. Also, a stiffer pipe may be required (see the section "Wall buckling" in Chap. 3).

Temperature and longitudinal stresses. Welded steel lines are subject to high-temperature induced and other longitudinal stresses. Expansion joints and/or closure welds will reduce these stresses and may be required. AWWA M-11 and AWWA C-206 make specific recommendations concerning expansion joint and closure welds.

Fiber reinforced plastic. Reinforced thermosetting resin pipes are widely used in the industrial market but have gained very little acceptance in the U.S. public works market. Aside from the reinforcing aspect, such as fiberglass, the primary difference in these materials is the fact that thermosetting resins cannot be melted and reformed while thermoplastic resin can. Members of the thermosetting plastic family include epoxy, polyester, and phenolic resins.

ASTM D2996 is the standard specification for filament-wound reinforced thermosetting resin pipe. ASTM D2997 is a standard specification for centrifugally cast reinforced thermosetting resin pipe. Within these standards, there are pipe designation codes and definitions.

ASTM D2310 is the standard classification for machine-made reinforced thermosetting resin pipe. This standard contains a complete definition of the various class and types of RTR pipes. It includes the following:

Manufacturing process

Type 1: Filament wound

Type 2: Centrifugally cast

Type 3: Pressure laminated

Resin used

Grade 1: Glass-fiber reinforced—epoxy

Grade 2: Glass-fiber reinforced—polyester

Grade 3: Glass-fiber reinforced—phenolic

Grade 4: Asbestos reinforced—polyester

Grade 5: Asbestos reinforced—epoxy

Grade 6: Asbestos reinforced—phenolic

Liner classification

Class A: No liner

Class B: Polyester resin—nonreinforced

Class C: Epoxy resin—nonreinforced

Class D: Phenolic resin—nonreinforced

Class E: Polyester resin—reinforced

Class F: Epoxy resin—reinforced

Class G: Phenolic resin—reinforced

Class H: Thermoplastic resin

ASTM D2992 is the standard method for obtaining hydrostatic-design basis for reinforced thermosetting resin pipe and fittings. There are two procedures:

Procedure A: cyclic strength.   This procedure is based on pipe failure at a minimum of 150×10^6 cycles at 25 cycles/min, 11.4 yr.

Procedure B: static strength.   This procedure is based on pipe failure at a minimum 100,000 hr (11.4 yr) of static pressure.

It is important to note that ASTM specifications for RTR pipe do not specify a service factor. Therefore, it is up to the design engineer to determine the hydrostatic design basis to be used for a particular pipe.

The product designation code, the manufacturer's product data and ASTM standards make it easy to determine what safety factor is being employed at the recommended working pressure.

The hydrostatic-design bases are listed in the applicable ASTM specifications. In the case of ASTM D2996 for filament wound RTR pipe, the following hydrostatic-design bases categories are listed:

Cyclic test method		Static test method	
Designation	Hoop stress, lb/in^2	Designation	Hoop stress, lb/in^2
A	2,500	Q	5,000
B	3,150	R	6,300
C	4,000	S	8,000
D	5,000	T	10,000
E	6,300	U	12,500
F	8,000	W	16,000
G	10,000	X	20,000
H	12,000	Y	25,000
		Z	31,500

Equation (4.3) can be utilized to calculate pressure ratings for RTR pipe.

An AWWA standard for glass-fiber reinforced, thermosetting resin pressure pipe, AWWA C-950, was approved in 1981; it incorporates information from these ASTM standards.

AWWA C-950 can be summarized as follows:

Manufacturing processes.

Type I filament wound

Type II centrifugally cast

Resins

Epoxy and polyester for RTRP and RPMP construction

Liners

None

Thermoplastic

Reinforced thermoset

Nonreinforced thermoset

Size range: 1 in to 144 in

Diameters

Inside diameters

Outside diameters (IPS)

Outside diameter (CI)

Metric dimensions

Pressure classes

50 lb/in², 100 lb/in², 150 lb/in², 200 lb/in², 250 lb/in², and over 250 lb/in²

Hydro safety factors: service and distribution pipe

2 to 1 (including surge)

Transmission pipe 1.4 to 1 (including surge)

A combined loading analysis is outlined in Appendix A of AWWA C-950. Two methods are given for calculating the combined stresses and strains. One method has been attributed to Spangler and the other to Molin. Both equations are to be used and the lowest value is the resulting combined stress and/or strain. This lowest value must not exceed the long-term bending strength of the product reduced by a design factor.

Research results from Utah State University indicate that this procedure is conservative. Figure 4.10 shows a typical plot of the AWWA C-950 combined strains versus internal pressure. The dashed line is typical pipe behavior as observed in all data sets from actual tests. The correla-

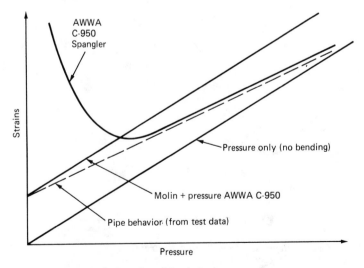

Figure 4.10 Typical plot of combined strains.

tion between test data and the AWWA C-950 method is generally acceptable. However, at low to intermediate pressure, particularly in the region where the Spangler and Molin curves cross, there is some discrepancy. The error is such that the suggested method is on the safe side.

Thrust restraint

Unbalanced hydrostatic and hydrodynamic forces in piping systems are called thrust forces. In the range of pressures and fluid velocities found in waterworks or wastewater piping, the hydrodynamic thrust forces are generally insignificant in relation to the hydrostatic thrust forces and are usually ignored. Simply stated, thrust forces occur at any point in the piping system where the direction or cross-sectional area of the waterway changes. Thus there will be thrust forces at bends, reducers, offsets, tees, wyes, dead ends, and valves.

Balancing thrust forces in underground pipelines is usually accomplished with bearing or gravity thrust blocks, restrained joint systems, or combinations of these methods. The internal hydrostatic pressure acts perpendicularly on any plane with a force equal to the pressure (P) times the area (A) of the plane. All components of these forces, acting radially within a pipe, are balanced by circumferential tension in the wall of the pipe. Axial components acting on a plane perpendicular to the pipe through a straight section of the pipe are balanced internally by the force acting on each side of the plane. Consider, however, the case of a bend as shown in Fig. 4.11.

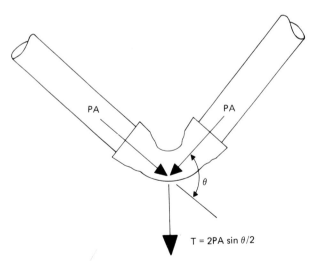

Figure 4.11   Thrust force. (*Reprinted from Thrust Restraint Design for Ductile Iron Pipe, by permission of the Ductile Iron Pipe Research Association.*)

The forces PA acting axially along each leg of the bend are not balanced. The vector sum of these forces is shown as T. This is the thrust force. In order to prevent separation of the joints, a reaction equal to and in the opposite direction of T must be established.

Figure 4.12 depicts the net thrust force at various other configurations. In each case, the expression for T can be derived by the vector addition of the axial forces.

Thrust blocks. For buried pipelines, thrust restraint is achieved by transferring the thrust force to the soil structure outside the pipe. The objective of the design is to distribute the thrust forces to the soil structure in such a manner that joint separation will not occur in unrestrained joints.

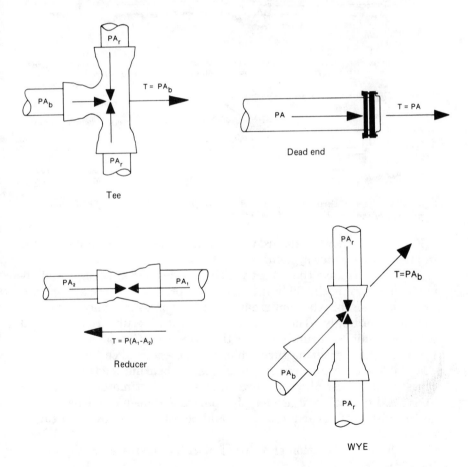

Figure 4.12 Thrust forces. (*Reprinted from Thrust Restraint design for Ductile Iron Pipe, by permission of the Ductile Iron Pipe Research Association.*)

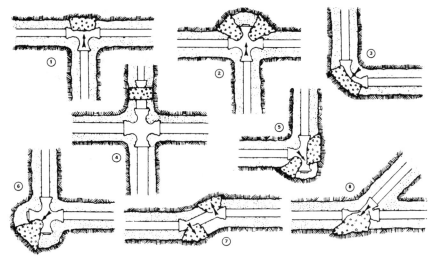

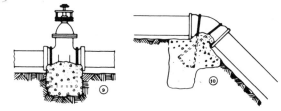

1. *Thru line connection, tee*
If thrusts, due to high pressure, are expected, anchor valves as
below. At vertical bends, anchor to resist outward thrusts.
2. *Thru line connection, cross used as tee*
3. *Direction change, elbow*
4. *Change line size, reducer*
5. *Direction change, tee used as elbow*
6. *Direction change, cross used as elbow*
7. *Direction change*
8. *Thru line connection, wye*
9. *Valve anchor*
10. *Direction change vertical, bend anchor*

Figure 4.13   Types of thrust blocking.   (*Reprinted from the Handbook of PVC Pipe,*[15] *by permission of the Uni-Bell PVC Pipe Association.*)

Figure 4.13 shows standard types of thrust blocking commonly used in pressurized water systems.

Table 4.7 displays the thrust which may develop at fittings and appurtenances for each 100 lb/in^2 of internal pressure. These are approximate values. Thrusts from greater or lesser pressures may be proportioned accordingly. The largest thrust may result from the test pressure, which is usually higher than the operating pressure.

One method for sizing thrust blocks uses assumed soil bearing values. Table 4.8 gives approximate allowable bearing loads for various types of soil. These allowable bearing loads are estimates only, are for horizontal thrusts, and are for pipe buried 2 ft deep or deeper. When doubt exists, safe bearing loads should be established by soil bearing tests.

The design calculation of a thrust block is illustrated in the following example:

TABLE 4.7   Thrust Developed per 100 lb/in² Pressure

Pipe size, in	Fitting 90° elbow, lbf	Fitting 45° elbow, lbf	Valve tees dead ends, lbf
4	2,560	1,390	1,810
6	5,290	2,860	3,740
8	9,100	4,920	6,430
10	13,680	7,410	9,680
12	19,350	10,470	13,690
14	26,010	14,090	18,390
16	33,640	18,230	23,780
18	42,250	22,890	29,860
20	51,840	28,090	36,640
24	73,950	40,070	52,280
30	113,770	61,640	80,420
36	162,970	88,310	115,210

TABLE 4.8   Estimated Bearing Load

Soil type	lb/ft²
Muck, peat, etc.	0
Soft clay	500
Sand	1000
Sand and gravel	1500
Sand and gravel with clay	2000
Sand and gravel cemented with clay	4000
Hard pan	5000

Example   Required, thrust block at 10-in 90° elbow. Maximum test pressure is 200 lb/in². Soil type: Sand and gravel with clay.

- Calculate thrust. From Table 4.7, thrust on 10-in 90° elbow = 13,680 lb per 100 lb/in² operating pressure; total thrust = 2 (13,680) = 27,360 lb.
- Calculate thrust block size. From Table 4.8, safe bearing load for sand-gravel-clay = 2000 lb/ft²; total thrust support area = 27,360/2000 = 13.68 ft².
- Select type of thrust block. From Figure 4.10, select type 3.

Restrained joints.   An alternate method of thrust restraint is with the use of restrained joints. Various mechanical locking-type joints are available to provide longitudinal restraint. Of course, a welded steel joint is considered to be rigid and provides maximum longitudinal restraints.

Restrained joint systems are subjected to the same thrust forces, but these forces are resisted or distributed over the restrained pipe length. The necessary length of restrained pipe, interacting with the soil, may be determined by the design engineer. Referring to Fig. 4.14, the restrained length on each side of the joint is L. The frictional resistance

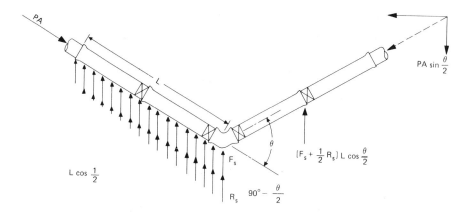

Figure 4.14 Free-body diagram for pipe with restrained joints. (*Reprinted from Thrust Restraint Design for Ductile Iron Pipe, by permission of the Ductile Iron Pipe Research Association.*)

and bearing resistance are given by F_s and R_b respectively. Summation of forces results in the following:

$$PA \sin \frac{\theta}{2} = F_s L \cos \frac{\theta}{2} + \frac{1}{2} R_b L \cos \frac{\theta}{2}$$

or

$$L = \frac{(PA \tan \theta/2)}{(F_s + \frac{1}{2} R_b)}$$

where P = internal pressure
A = pipe cross-sectional area
F_s = frictional force
F_b = bearing force

For a cohesionless soil, the friction force F_s may be calculated as follows:

$$F_s = W \tan \gamma$$

where $W = 2W_e + W_p$
$\gamma = f_\phi \phi$
W_e = total soil load
W_p = weight of pipe plus water
f_ϕ = friction factor between pipe and soil
ϕ = internal friction angle of soil

The above method will generally produce conservative results. If cohesion is present, cohesive forces will also be involved which will make results even more conservative. However, since cohesive forces are time dependent, it is recommended that they be neglected.

Safety factors

Design of pressure pipe is based upon certain performance limits such as long-term hydrostatic burst pressure and/or crush load either acting independently or simultaneously. The allowable total stress or strain is equal to the failure stress or strain reduced by a safety factor. For example:

$$\sigma_A = \frac{\sigma_f}{SF} \quad \text{or} \quad \epsilon_A = \frac{\epsilon_f}{SF}$$

where σ_A = allowable stress
σ_f = failure stress
ϵ_A = allowable strain
ϵ_f = failure strain
SF = safety factor

The total working stress/(strain) must be equal to or less than the allowable stress/(strain). If a combined loading analysis is not required, stresses due to internal pressure and external loads are evaluated separately, and the safety factor is applied to the largest value. For combined loading, the safety factor is applied to the combined stress.

For nonlinear failure theories such as the Schlick formula, safety factors must be applied to both internal pressure and external load. These two factors of safety need not be equal.

For plastic pipe, the design is based on life rather than a failure stress. As previously discussed in this chapter, a hydrostatic-design basis (stress) is established on the basis of a life of 100,000 hr. The design stress is this hydrostatic-design basis reduced by a factor of safety. A factor of safety of 2.0 will give, essentially, infinite life since the stress regression curve is linear on a log-log plot (see Fig. 4.6).

Standards for each pipe product may list recommended safety factors. Also, manufacturers often recommend certain safety factors for their products. The bases for the calculations of these are often quite different. The design engineer should be aware of these differences when comparing products and should always have the option of requiring a safety factor which is different from the recommended value. The need for safety factors arises mainly from uncertainties. These uncertainties range from pipe manufacture to pipe installation conditions. The more the uncertainty, the higher the safety factor should be. The engineer should be very cautious in utilizing safety factors that are lower than those recommended by national standards or by the manufacturer.

Bibliography

1. Andrews, James S., "Water Hammer Generated during Pipeline Filling," masters thesis, Colorado State University, Fort Collins, Colo., August 1970.
2. ASTM, "Standard Method of Test for Time-To-Failure of Plastic Pipe Under Long-Term Hydrostatic Pressure—ASTM D1598," American Society of Testing and Materials, Philadelphia, Pa., 1976.
3. AWWA standards: M-11, M-9, M-23, C-150, C-200, C-206, C-300, C-301, C-303, C-400, C-401, C-402, C-403, C-900, C-901, and C-950, American Water Works Association, Denver, Colo.
4. Bishop, R. R., "Course Notebook," Utah State University, Logan, Utah, 1983.
5. Devine, Miles, "Course Notebook," Utah State University, Logan, Utah, 1980.
6. Hucks, Robert T., "Design of PVC Water Distribution Pipe," Civil Engineering ASCE, vol. 42, no. 6, June 1972, pp. 70–73.
7. Jeppson, Roland W., Gordon H. Flammer, Gary Z. Watters, "Experimental Study of Water Hammer in Buried PVC and Permastran® Pipes," Utah Water Research Laboratory/College of Engineering, Utah State University, Logan, Utah, April 1972.
8. Jeppson, Roland W., Gordon H. Flammer, and Gary Z. Watters, "Experimental Study of Water Hammer in Buried PVC and PERMASTRAN Pipes," Utah Water Research Laboratory, PRWG0113-1, March 1972.
9. Kerr, S. Logan, "Water Hammer-a Problem in Engineering Design," Consulting Engineering, May 1985.
10. Lamé, G., "Lecons sur la théorie de l'élasticité," Paris, 1852.
11. Moser, A. P., "Course Notebook," Utah State University, Logan, Utah, 1983.
12. Moser, A. P., John Clark, and D. P. Bair, "Strains Induced by Combined Loading in Buried Pressurized Fiberglass Pipe," Proc. ASCE International Conference on Advances in Underground Pipeline Engineering, ASCE, Madison, Wis. (1985).
13. Streeter, Victor L., Fluid Mechanics, 2d ed., McGraw-Hill, New York, 1958, pp. 175–187.
14. Ductile Iron Pipe Research Association, "Thrust Restraint Design for Ductile Iron Pipe," Birmingham, Ala., 1984.
15. Uni-Bell PVC Pipe Association, Handbook of PVC Pipe, 2d ed., Uni-Bell, Dallas, 1982
16. Vinson, H. W., "Response of PVC Pipe to Large, Repetitive Pressure Surges," Proc. of the International Conference on Underground Plastic Pipe, American Society of Civil Engineers, New York, March 1981.
17. Walker, Robert P., "Course Notebook," Utah State University Logan, Utah, 1983.
18. Watters, G. Z., "The Behavior of PVC Pipe Under the Action of Water Hammer Pressure Waves," Utah Water Research Laboratory, PRWG-93, March 1971.

Chapter

5

Buried Pipe Products

Chapter 5 deals with various generic pipe products in two general classes; rigid pipe and flexible pipe. For each product, selected standards and material properties are listed. The standards are from standard organizations such as the American Water Works Association (AWWA) and American Society for Testing and Materials (ASTM). Actual design examples for the various products are given in this chapter.

Rigid Pipes

Asbestos-cement pipe

Asbestos-cement (AC) pipes are available for both gravity and pressure applications (see Tables 5.1 and 5.2). However, in recent years production of this product in the United States has been limited to

TABLE 5.1 Properties and Design Constants

Modulus of elasticity	3.0×10^6 lb/in^2
Tensile strength	3000–4000 lb/in^2
Shear strength	4000 lb/in^2 across pipe axis
Modulus of rupture (MR)	5000–6000 lb/in^2 (bending strength in crush)
Compressive strength	7000 lb/in^2
Thermal conductivity	$K = 5.5$ (Btu · in)/(h · °F · ft^2), 4 when perfectly dry
Thermal coefficient of expansion	$4-5 \times 10^{-6}$ in/in/°F
Specific heat	0.27 Btu/lb°F @ 212°F
Moisture coefficient of expansion	$1.5-2.0 \times 10^{-5}$ in/in/(% moisture change) (moisture content is 6 to 7% for normal atmospheric conditions and 15 to 20% for fully saturated conditions)
Hazen-Williams coefficient	$C = 140$
Manning's coefficient	$n = 0.010$

TABLE 5.2 Applicable National Standards

AWWA C-400	Asbestos-cement distribution pipe, 4 in through 16 in
AWWA C-401	Standard practice for the selection of asbestos-cement distribution pipe
AWWA C-402	Standard for asbestos-cement transmission pipe 18 in through 42 in
AWWA C-403	Standard practice for the selection of asbestos-cement transmission and feeder main pipe, sizes 18 in through 42 in
AWWA C-603	Standard for installation of asbestos-cement water pipe
ASTM C-296	Asbestos-cement pressure pipe
ASTM C-428	Asbestos-cement nonpressure sewer pipe
ASTM C-460	Standard definitions of terms relating to asbestos-cement and related products
ASTM C-500	Standard methods of testing asbestos-cement pipe
ASTM C-663	Asbestos-cement transmission pipes
ASTM D-1869	Specification for rubber rings for asbestos-cement water pipe
Federal Specification SS-P-351c	Specification for pipe, asbestos-cement for underwater pressure
Bureau of Yards and Docks, Navdocks, DM-5	U.S. Navy Civil Engineering Design Manual (includes asbestos-cement pressure pipe)
Canadian Standards Association	
CSA B127.1	Components for use in AC DWV systems
CSA B122.2	Components for use in AC building sewer systems
CSA B127.11	Recommended Practice for the Installation of AC DWV pipe and fittings

pipe for pressure applications only. Because of the health risks associated with the handling of asbestos, AC pipe productions in the United States may come to a complete halt in the near future. This product has some flexibility especially for lower classes (thinner walls). However it is generally considered to be a rigid pipe product and the rigid pipe design method should be used for AC pipe installations.

Asbestos-cement pipes are manufactured from asbestos, cement, silica, and water. The pipe-making machine places this mixture on a polished steel mandrel and it is pressure-steam treated (autoclaved) to achieve curing with less than 1 percent uncombined calcium hydroxide (free lime). AC pipe which have less than 1 percent free lime are designated as type 2. Type 2 AC pipes are generally resistant to all levels of soluble sulfates, but will be attacked by acids with a pH level of 5.0 or less. Type 1 pipes have more than 1 percent free lime and are generally not resistant to either soluble sulfates or to acids.

Asbestos cement pipes are joined via rubber gasketed couplings.

The pipe has a hard and fairly smooth internal surface. A Hazen Williams coefficient of 140 and a Manning coefficient of 0.010 should be used for long-term design. AC pipe can be tapped for water service. Also, various fittings are available for connections. Field cutting for repairs and/or installation is possible. Appropriate safety precautions should be followed to protect workers from air-borne asbestos dust. Manufacturer's safety procedures should be followed.

The asbestos-cement-silica composite achieves a remarkably high tensile strength of up to 4000 lb/in^2. This high strength is directly attributable to the asbestos fiber reinforcement. AC pipe is durable as many AC pipelines have been operating for more than 35 years. AC pipes are not as prone to impact damage as some rigid pipes, nevertheless care should be taken in handling. When excavating, to make connections to or in repairing AC pipes, care must be taken to prevent the backhoe bucket from damaging the line. For water systems, AC pipes are available for both transmission and distribution systems. It is also available for various specialty applications.

Example 5.1 (Gravity Storm Sewer)   A 36-in diameter storm sewer line is to be installed. It passes through a small hill which requires a trench 20 ft deep. Native material, a silty sand with clay, will be used for final backfill. The trench width at the top of the pipe is not to exceed 7 ft. Calculate the minimum strength of asbestos-cement pipe for both "B" and "C" bedding. Also, what strength will be required for a possible "worst case" if the trench width exceeds the transition width and only "C" bedding is achieved. Ground water is 10 ft below surface. (See Chap. 2 for design criteria.)

1. Determine earth load (ditch condition)

$$\frac{H}{B_d} = \frac{20}{7} = 2.86 \qquad [\text{from Fig. 2.2 } (C_d = 1.9)]$$

Clayey sand, $K\mu = 0.150$

$$W_d = C_d\gamma B_d^2 = 1.9\,(120\ lb/ft^3)\,(7)^2 = 11,172\ lb/ft$$

where γ = unit weight of soil (assume 120 lb/ft^3)
B_d = trench width at top of pipe

2. Determine earth load (if transition width exceeded)

$$\frac{H}{B_c} = \frac{20}{3} = 6.67 \qquad \left[\text{From Fig. 2.6}\left(\frac{B_d}{B_c} = 2.65\right)\right]$$

Assume $r_{sd}p = 0.75$

$$B_d\ (\text{transition}) = \left(\frac{B_d}{B_c}\right)B_c = 2.65(3) = 7.95\ feet \approx 8.0$$

$$W_d\ (\text{transition}) = C_d\,\gamma\,B_d^2 = 1.9\,(120)\,(8)^2 = 14,592\ lb/ft$$

3. Determine class of pipe required for "B" bedding, 7-foot trench width

Load factor (LF) for "B" bedding = 1.9   (see Table 3.2)

$$LF = 1.9 \quad \text{(see Table 3.2)}$$

$$\text{Safety factor} = 1.5$$

$$\text{Required strength} = \text{(design load)} \left(\frac{\text{safety factor}}{\text{load factor}}\right)$$

$$= W_d\left(\frac{SF}{LF}\right) = \frac{11,172\,(1.5)}{1.9} = 8,200 \text{ lb/ft}$$

Contact manufacturer to see if this strength or higher is available.

4. Determine strength of pipe required for "C" bedding, 7-foot trench width

$$\text{Load factor (LF) for "C" bedding} = 1.5 \quad \text{(see Table 3.2)}$$

$$\text{Safety factor} = 1.5$$

$$\text{Required strength} = W_d \left(\frac{SF}{LF}\right) = \frac{11,172(1.5)}{(1.5)}$$

$$= 11,172 \text{ lb/ft}$$

Again, contact manufacturer for the availability of this strength. Although this is a nonpressure application, a pressure pipe with the required crush strength may be used.

5. Determine strength of pipe required if transition width is reached or exceeded, class "C" bedding. From item 2 above

$$\text{load } W_d = 14,592 \text{ lb/ft}$$

$$\text{Required strength} = W_d\left(\frac{SF}{LF}\right) = \frac{14,592(1.5)}{1.5} = 14,592 \text{ lb/ft}$$

This strength or higher may *not* be available from manufacturer.

Asbestos-cement pressure pipes are designed using a combined loading theory (Schlick Formula) as discussed in Chap. 4. Equations (4.12) and (4.13) are repeated here for convenience.

$$w = W \sqrt{1 - \frac{p}{P}} \qquad (4.12) \qquad\qquad (5.1)$$

$$p = P\left[1 - \left(\frac{w}{W}\right)^2\right] \qquad (4.13) \qquad\qquad (5.2)$$

Maximum bending stress in a pipe subjected to three-edge loading can be calculated as follows:

$$\sigma = \frac{M(t/2)}{I} = \frac{M(t/2)}{b\,t^3/12} = \frac{(M/b)(t/2)}{t^3/12} \qquad (5.3)$$

$$M = 0.318\,F(r_i + t/2)* \qquad (5.4)$$

---

*This is the maximum moment in a closed ring loaded with diametrically opposite concentrated loads. See a text on mechanics of materials for details.

where M = moment
     F = load
     I = moment of inertia of wall
     r_i = internal radius
     b = length of specimen thickness
     t = wall thickness

Equation (5.3) can be written as follows:

$$\sigma = \frac{6(0.318)(F/b)(r_i + t/2)}{t^2} \tag{5.5}$$

For external loading only, at failure, the stress σ is the strength sometimes called the modulus of rupture (MR). F/b is the three-edge bearing load to cause failure (three-edge bearing strength W) in lb/ft. Thus,

$$MR = \sigma = \frac{6(0.318)(W\text{lb/ft})(1\text{ft}/12\text{in})[(d_i + t)/2]}{t^2}$$

or

$$MR = \frac{0.0295(W)(D + t)}{t^2} \tag{5.6}$$

The hoop stress σ_h in a cylinder may be calculated as follows:

$$\sigma_h = \frac{PD}{2t} \tag{5.7}$$

Knowledge of σ_h, MR and t will allow calculation of w and p through the use of Eqs. (5.1) or (5.2) and (5.6) and (5.7). By solving Eq. (5.7) for P and equation (5.6) for W, one may substitute into Eqs. (5.1) and (5.2) to obtain

$$w = \frac{MRt^2}{0.0795\,(D + t)} \sqrt{\frac{\sigma 2t/D - p}{\sigma 2t/D}} \tag{5.8}$$

and

$$p = \frac{\sigma 2t}{D}\left\{1 - \left[\frac{w}{MRt^2/(D + t)0.0795}\right]^2\right\} \tag{5.9}$$

respectively.

Equations (5.8) and (5.9) above express the external and internal loads, for a pipe of given thickness, modulus of rupture, and tensile strength, which will cause failure when applied simultaneously. It is difficult to solve these equations explicitly. The general procedure is to construct a graphical solution for standard pipe classifications and standard installation and operating conditions (see AWWA C-402). In addition, the design process will require the application of an appropriate factor of safety.

Thickness design of asbestos-cement class pressure pipe is outlined in AWWA C-401. Two cases of design are considered. The design methods are summarized in Fig. 5.1. Note that the safety factors recommended are different. Asbestos-cement pipe designed by considering case I will generally exceed the capability of pipe designed by case II.

The nature of transmission systems has been recognized by AWWA. AWWA C-402-75 covers a wide range of pipe classifications suited to provide exactly the right pipe for the design conditions encountered. In cases where operating and installation conditions are controlled and the magnitude of potential surge pressures are known, lower safety factors may be justified. Figure 5.1 also summarizes such a design procedure for asbestos-cement transmission pipe, case III.

Example 5.2 (Distribution Line)   A 12-in diameter distribution line will operate at a working pressure of 100 lb/in^2. Average depth of cover will be 5.0 ft under a paved roadway. The native soil is sand. Using standard AWWA design procedures, what class asbestos-cement should be used if the pipe is laid in a flat bottom trench with backfill tamped. Assume the trench width is 3.0 ft., and the bedding factor is 1.3.

Design Case	Internal Load Design	External Load Design
Case I (live load is zero)	p = (operating pressure) × 4.0 SF = 4.0	w = (transition load) × 2.5 SF = 2.5
Case II (surge pressure is zero)	p = (operating pressure) × 2.5 SF = 2.5	w = (earth + live load) × 2.5 SF = 2.5
Case III (transmission) (designed for specific surge pressure)	p = (operating pressure + surge pressure) × 2.5 SF = 2.5	w = (earth + live load) × 2.5 SF = 2.5

Figure 5.1   Asbestos-cement pressure pipe design summary (See AWWA C401).

Case I:

Hydrostatic: $P = \text{class} \times SF$ $SF = 4.0$

Crush load: $W = \left[\dfrac{\text{earth load (transition)}}{\text{bedding factor}}\right]SF$

$SF = 2.5$

Case II:

Hydrostatic: $P = \text{class} \times SF$ $SF = 2.5$

Crush load: $W = \left[\dfrac{\text{(earth load + live load + impact)}}{\text{bedding factor}}\right]SF$

$SF = 2.5$

In both case I and II use combined loading

$$w = W\sqrt{\dfrac{P - p}{P}}$$

Figure 5.2 Outline of design procedure for asbestos-cement class pressure pipe (6 to 16 in).

Hydrostatic design:

$P = (\text{operating pressure + surge pressure})SF$

$SF = 2.0$

Crush design:

$W = [\text{earth load (transition) + live load (H=20) bedding factor}]SF$

$SF = 1.5$

$$w = W\sqrt{\dfrac{P - p}{P}}$$

Figure 5.3 Outline of asbestos-cement transmission pipe design procedure (18–36 in)

1. Determine earth load

$$\frac{H}{Bd} = \frac{5.0}{3.0} = 1.67 \quad (\text{From Fig. 2.2, } C_d = 1.3)$$

$$K\mu = 0.165 \quad (\text{for sand})$$

$$W_d = 1.3\,(120)(3.0)^2 = 1404 \text{ lb/ft}$$

2. Determine load at transition width*

$$\frac{H}{B_c} = \frac{5.0}{(13.74/12)} = 4.37 \quad (\text{From Fig. 2.6 with } r_{sd}P = 0.5 \text{ and } K\mu$$

$$= 0.165,\ B_d/B_c = 2.22)$$

$$B_d(\text{transition}) = \frac{13.74}{12}\,2.22 = 2.54 \text{ ft}$$

---

*13.74 is OD which can be obtained from manufacturers' specifications.

$$\frac{H}{B_d} = \frac{5.0}{2.54} = 1.97 \qquad \text{(From Fig. 2.2, } C_d = 1.4)$$

$$K\mu = 0.165$$

$$\text{Load is } W_d = 1.4(120)(2.54)^2 = 1084 \text{ lb/ft}$$

(not 1404 as determined previously). Alternately the load can be obtained using Fig. 2.5 for projecting conduits.

$$\frac{H}{B_c} = 4.37 \qquad \text{(From Fig. 2.5, } C_c = 7.0)$$

$$W_c = C_c \gamma B_c^2 = 7.0(120)\left(\frac{13.74}{12}\right)^2 = 1111 \text{ lb/ft}$$

The 1111 lb/ft is essentially the same as 1084 lb/ft as previously calculated. The error is due to graphical interpolations for C_d and C_c.

3. Determine live load

$$W_L = 340 \text{ lb} \qquad \text{(from Fig. 2.14)}$$

4. Determine total load

$$W_T = W_c + W_L = 1111 + 340 = 1450$$

5. Determine internal pressure requirement

Case I: (live load is zero)

$$p = (100)\, 4.0 = 400 \text{ lb/in}^2$$

$$w = \frac{1111\,(2.5)}{1.3} = 2136 \text{ lb/ft}$$

Case II: (surge is zero)

$$p = (100)\, 2.5 = 250 \text{ lb/in}^2$$

$$w = \frac{1450\,(2.5)}{1.3} = 2788 \text{ lb/ft}$$

By use of the Schlick formula we can now determine the P and W or wall thickness of the pipe which is required. For AC pipe, the performance criteria P and W may be solved by trial and error as follows (Class 100, $P = 490$ lb/in^2, $W = 5200$ lb/ft) see Fig. 5.4:

Case I:

$$w = W \sqrt{\frac{P - p}{P}} = 5200 \sqrt{\frac{490 - 400}{490}} = 2288 \text{ lb/ft}$$

$$2228 \geq 2136 \qquad \text{class 100 is acceptable}$$

Case II:

$$w = 5200 \sqrt{\frac{490 - 250}{490}} = 3639 \text{ lb/ft} > 2780 \text{ lb/ft}$$

Class 100 is acceptable.

Figure 14E of AWWA C-401-83 (see Fig. 5.4) is a solution for $r_{sd}P = 0.70$, $K\mu = 0.192$ and $\gamma = 120$ lb/ft^3. Class 100 pipe would

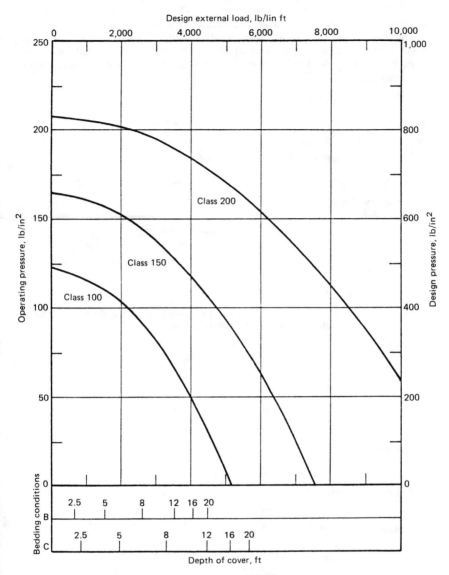

Figure 5.4 Class pipe-design curve for 12-in diameter asbestos cement pipe. (See AWWA C401 for other diameters). *(Reprinted, by permission, from ANSI/AWWA C401-83,[3] American Water Works Association, 1986.)*

therefore perform for any conditions which are below the lowest design curve. The class designations are based on case I with a class C bedding, excavated coupling holes, and 5.0 ft of cover.

Example 5.3 (Transmission Pipe Design)   A 24-in diameter transmission line will deliver water at 7000 gal/m and 5.0 ft/sec from a reservoir to a treatment plant 10 miles away. The pipe will be buried 5.0 ft deep in a 4.0-ft wide trench in sand carefully compacted or bedded with a coarse granular material up to the spring line. Surge devices and valve operating equipment will control maximum surge pressures to a maximum of 50 lb/in^2. The system will operate at maximum pressure of 150 lb/in^2. Determine the appropriate asbestos-cement transmission pipe. (See AWWA C-402.)

1. Determine earth load

$$\frac{H}{B_d} = \frac{5.0}{4.0} = 1.25 \qquad \text{(from Fig. 2.2, } C_d = 1.1)$$

$$K\mu = 0.165 \qquad \text{(for sand)}$$

$$W_d = 1.1 \, (120)(4.0)^2 = 2112 \text{ lb/ft}$$

Transition width

$$\frac{H}{B_c} = \frac{5.0}{(26.4/12)} = 2.27 \qquad \text{(from Fig. 2.6, } B_d/B_c = 1.8)$$

$$r_{sd}p = 0.5$$

$$B_d = 1.8 \, \frac{26.4}{12} = 3.96\text{ft,}$$

that is 4.0-ft trench width just exceeds the transition width and the load calculated for a 4.0-ft wide trench is just slightly conservative.

$$\frac{H}{B_c} = \frac{5.0}{2.2} = 2.27 \qquad \text{(from Fig. 2.5 } C_c = 3.4)$$

$$r_{sd}p = 0.5$$

$$W_c = 3.4 \, (120)(2.2)^2 = 1974 \text{ lb/ft}$$

$$W_d \approx W_c \, @ \text{ transition width}$$

2. Determine live load

$$W_L = 340 \qquad \text{(from Fig. 2.14)}$$

3. Determine total load

$$2112 + 340 = 2452 \text{ lb/ft}$$

4. Combined loading

$$p = (150 + 50) \, 2.0 = 400 \text{ lb/in/}^2$$

$$w = \left(\frac{2112 + 340}{1.9}\right)1.5 = 1936 \text{ lb/ft}$$

Try (T50, P = 500 lb/in^2, W = 8100 lb/ft

$$w = W \sqrt{\frac{P-p}{P}} = 8100 \sqrt{\frac{500-400}{500}} = 3622 \text{ lb/ft}$$

$$3622 \text{ lb/ft} > 1936 \text{ lb/ft} \qquad \text{T50 is acceptable}$$

Alternately the pipe could be chosen from the selection curves of Fig. 4.3 (see AWWA C-402-84).

Clay pipe

Vitrified clay pipe is manufactured from clays and shales which are chemically inert. In the manufacturing process, various clays and shales are pulverized and screened and placed in storage bins. Blended materials are carried to the pugmill and mixed and moistened with water for a proper mix consistency for extrusion. The mix is then forced through a die into a vacuum chamber where trapped air is removed. This mixture is then machine extruded in the form of pipe. This fresh extruded pipe contains about 18 percent water and is called greenware. Greenware is placed in drying rooms to reduce the moisture content to about 3 percent. The pipe is then taken to the kilns and preheated to approximately 400°F to drive off remaining moisture. The pipe travels slowly through the kiln reaching a temperature near 2000°F where vitrification takes place.

During vitrification the clay fuses into a very hard, chemically stable compound. Vitrified clay is very corrosion and abrasion resistant. Because of its inherent low strength, vitrified clay pipe is used for nonpressure applications only. It is brittle and subject to impact damage and special care in handling is a requirement.

Newer designs do not have extruded clay bells. Instead, a bell is formed by helically winding continuous glass filaments and a thermosetting resin to form a bell on a plain pipe end. A groove is molded into the bell for a rubber gasket.

Clay pipe is generally available in sizes ranging from 4-in diameter to 36-in diameter. However it may be available in some locations in diameters up to 42 in. The strength is determined by the three-edge bearing test, varies with diameter, and ranges from 2000 lb/ft to 7000 lb/ft (Tables 5.3 and 5.4).

Example 5.4   A 15-in diameter sanitary sewer line is to be installed 14 ft deep. Native material, which is sand, will be used for final backfill. If the trench

TABLE 5.3   Standards for Clay Pipe

ASTM C-700	Clay pipe, vitrified, extra strength, standard strength, and perforated
ASTM C-425	Compression joints for vitrified clay pipe and fittings
ASTM C-301	Clay pipe, vitrified (test methods)
ASTM C-12	Installing vitrified clay-pipe lines
ASTM C-828	Low-pressure air test of vitrified clay-pipe lines
Canadian Standards Association	
CSA A60.1	Vitrified clay pipe
CSA A60.2	Methods of testing vitrified clay pipe
CSA A60.3	Vitrified clay-pipe joints

TABLE 5.4   ASTM C700-74 Clay Pipe Minimum Crushing Strengths (Three-Edge Bearing Strength)

Nominal size, in	Extra strength lb/ft	Nominal size, in	Extra strength, lb/ft
3	2000	21	3850
4	2000	24	4400
6	2000	27	4700
8	2200	30	5000
10	2400	33	5500
12	2600	36	6000
15	2900	39	6600
18	3300	42	7000

width is 3.0 ft, what pipe and bedding classes should be selected? (Note: This example was previously given as Example 3.1).

1. Determine earth load

$$\frac{H}{B_d} = \frac{14}{3} = 4.67 \qquad \text{(from Fig. 2.2 } C_d = 2.4\text{)}$$

$$K\mu = 0.165 \qquad \text{(for sand)}$$

$$W_d = C_d\,\gamma\,B^2_d = 2.4\,(120)(3.0)^2 = 2592\ \text{lb/ft}$$

2. Determine live load

$$W_L = 150\ \text{lb/ft} \qquad \text{(from Fig. 2.14)}$$

Note: 150 ≪ 2592 (live load may be neglected)

3. Select bedding and load factor

Class D   LF = 1.1

Class C   (from Table 3.2)   LF = 1.5

Class B   LF = 1.9

4. Select pipe strength (safety factor = 1.5)

$$\text{Three-edge strength} = \frac{W_d(\text{SF})}{\text{LF}} = \frac{2592(1.5)}{\text{LF}}$$

Required Strength for Various Bedding Classes

Bedding class	LF	Three-edge, lb/ft	Required strength, lb/ft
B	1.9	2046	Extra strength (2900)
C	1.5	2592	Extra strength (2900)
D	1.1	3535	This strength is not available

Concrete pipe

Concrete pipe products are made by several processes. Included are nonreinforced products in sizes ranging from 4-in diameter to 36-in diameter and various reinforced products in sizes 12-in through 144-in diameter (see Tables 5.5, 5.6, and 5.7).

The nonpressure types are described in ASTM C-14 for nonreinforced and in ASTM C-76 for the reinforced-type, and in CSA A-257 for both types.

Concrete pressure pipe includes various types of wall construction. Some are designed and manufactured for specific service applications and other types are constructed to be suitable for a broad range of applications.

Prestressed concrete cylinder pipe   Prestressed concrete cylinder pipe has two types of construction: embedded-cylinder types and the lined-cylinder types (Figs. 5.5 and 5.6). In both types, manufacturing begins with a welded steel cylinder to which joint rings are attached to each end. This steel cylinder is then hydrostatically tested.

TABLE 5.5   ASTM C-14 Nonreinforced Concrete Pipe

Pipe diameter, in	Minimum strength in three-edge bearing, lb/ft		
	Class 1	Class 2	Class 3
4	1500	2000	2400
6	1500	2000	2400
8	1500	2000	2400
10	1600	2000	2400
12	1800	2250	2600
15	2000	2600	2900
18	2200	3000	3300
21	2400	3300	3850
24	2600	3600	4400
27	2800	3950	4600
30	3000	4300	4750
33	3150	4400	4875
36	3300	4500	5000

TABLE 5.6   ASTM C-76 Reinforced Concrete Pipe *D* Load (lb/ft · ft dia.) Required

Class	Size range, in dia.	0.01 in crack	Ultimate
I	60–144	800	1200
II	12–144	1000	1500
III	12–144	1350	2000
IV	12–144	2000	3000
V	12–144	3000	3750

TABLE 5.7   AWWA and ASTM Standards for Concrete Pipe

AWWA C-300	Standard for reinforced concrete pressure pipe, steel cylinder-type for water and other liquids
AWWA C-301	Standard for prestressed concrete pressure pipe, steel cylinder-type for water and other liquids
AWWA C-302	Standard for reinforced concrete pressure pipe, noncylinder-type for water and other liquids
AWWA C-303	Standard for reinforced concrete pressure pipe, steel cylinder-type, pretensioned for water and other liquids
AWWA Manual 9	Concrete pressure pipe, manual of water supply practices
ASTM C-118	Concrete pipe for irrigation or drainage
ASTM C-14	Concrete sewer, storm-drain, and culvert pipe
ASTM C-505	Nonreinforced concrete irrigation pipe with rubber-gasket joints
ASTM C-985	Nonreinforced concrete specified strength culvert, storm-drain, and sewer pipe
ASTM C-654	Porous concrete pipe
ASTM C-506	Reinforced concrete arch culvert storm-drain and sewer pipe
ASTM C-76	Reinforced concrete culvert, storm-drain, and sewer pipe
ASTM C-655	Reinforced concrete D-load culvert, storm-drain, and sewer pipe
ASTM C-507	Reinforced concrete elliptical culvert, storm-drain, and sewer pipe
ASTM C-361	Reinforced concrete low-pressure pipe
ASTM C-924	Low-pressure air test of concrete pipe sewer lines

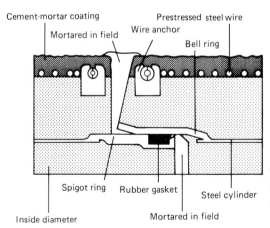

Figure 5.5   Wall cross-section of embedded cylinder pipe. *(Reprinted from Bulletin #200, Embedded Cylinder, by permission of the United Concrete Pipe Corporation.)*

Cement-mortar coating

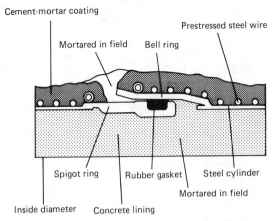

Prestressed steel wire

Mortared in field   Bell ring

Spigot ring   Rubber gasket   Steel cylinder

Mortared in field

Inside diameter   Concrete lining

Figure 5.6 Wall cross-section of lined cylinder pipe. *(Reprinted from Bulletin #200, Line Cylinder, by permission of the United Concrete Pipe Corporation.)*

A concrete core is either cast (embedded-cylinder) or spun (lined-cylinder) in the steel cylinder. After curing, the cylinder is helically wrapped with hard-drawn wire under high-tensile stress. The lead angle is controlled to produce a specific compression stress in the concrete core. After wrapping, the pipe is coated with a cement slurry and a dense mortar or concrete coating.

Embedded cylinder pipe is commonly available in sizes 24-in through 144-in diameter. Lined-cylinder pipe is manufactured in diameters 16 in through 60 in. Prestressed concrete cylinder pipe is designed using a combined loading analysis. This method was discussed in Chap. 4 (see also AWWA C-301).

Reinforced concrete cylinder pipe   This pipe is similar to the embedded-cylinder pipe in manufacture. However, no prestressed wire is applied and instead one or more reinforcing cages and the steel cylinder are positioned between vertical forms and the concrete is cast (Fig. 5.7). Steam or water is used to cure the concrete. This pipe is available in diameters of 24 in through 144 in. Design is based on either the strength method or the working stress method. In either cases, the pipe is to be designed to withstand internal pressure and external load, each acting separately or in combination (see AWWA C-300, Appendix A).

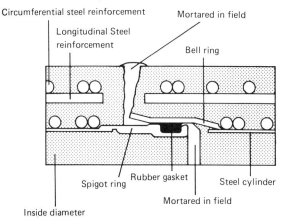

Circumferential steel reinforcement   Mortared in field

Longitudinal Steel reinforcement   Bell ring

Spigot ring   Rubber gasket   Steel cylinder

Mortared in field

Inside diameter

Figure 5.7   Wall cross-section of reinforced concrete cylinder pipe. *(Reprinted from Bulletin #200 Reinforced Concrete Cylinder Pipe, by permission of the United Concrete Pipe Corporation.)*

Reinforced concrete noncylinder pipe   This type of concrete pipe is manufactured by positioning one or more steel cages in proper radial location(s) (Fig. 5.8). The cages are placed between two vertical forms and the concrete is cast. Alternately, the cages are attached to an outer form and the entire assembly is rotated and the concrete is cast centrifugally. AWWA C-302 outlines a design procedure for internal pressure and external loads acting simultaneously. Reinforced noncylinder pipe is available in diameters of 12 in through 144 in.

Pretensioned concrete cylinder pipe   In the manufacture of pretensioned concrete cylinder pipe, one starts with steel cylinders made from steel coils and spirally welded or made from steel sheet and welded longitudinally. End rings are welded to the steel cylinder and it is then hydrostatically tested to 75 percent of yield strength of the steel. A cement mortar lining is applied centrifugally. After curing, the cement-mortar line-steel cylinder is pretensioned by helically winding steel rod under a small tension to the outside of the steel cylinder. The pitch of the winding is controlled by specific design requirements. A cement-mortar coating is then applied to the exterior surface of the rod-wrapped cylinder and the completed pipe is cured (**Fig. 5.9**). This pipe is normally available in diameters 10 in through 42 in. The design of this pipe is based on an analysis of both internal pressure and external loads acting separately but not in combination. This design method is usually used for flexible pipe which pretensioned concrete is not. The pipe must be installed in such a manner that the deflection is less than $D_2/4000$ (see AWWA C-303, Appendix A).

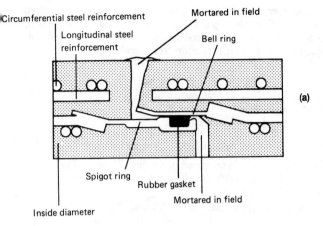

Circumferential steel reinforcement

Longitudinal steel reinforcement

Mortared in field

Bell ring

(a)

Spigot ring

Rubber gasket

Mortared in field

Inside diameter

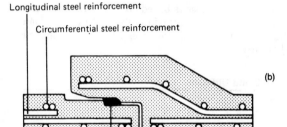

Longitudinal steel reinforcement

Circumferential steel reinforcement

(b)

Inside diameter

Rubber gasket

Figure 5.8 Wall cross-section of reinforced concrete noncylinder pipe (a) (with steel joint rings): (b) (with concrete bell and spigot). *Reprinted from Bulletin #200, Reinforced Concrete Cylinder Pipe, by permission of the United Concrete Pipe Corporation.)*

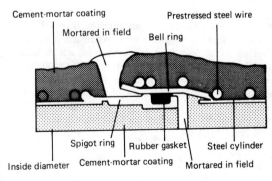

Cement-mortar coating

Prestressed steel wire

Mortared in field

Bell ring

Spigot ring

Rubber gasket

Steel cylinder

Inside diameter   Cement-mortar coating   Mortared in field

Figure 5.9 Wall cross-section of pretensioned concrete cylinder pipe. *(Reprinted from Bulletin #200, Pretensioned Concrete Shot-Cote Pretensioned Pipe, by permission of the United Concrete Pipe Corporation.)*

Example 5.5 A 15-in diameter sanitary sewer line is to be installed 14 ft deep in native sand. The trench width at the top of the pipe is to be 3.0 ft. For class B, class C, and class D bedding, select the required strength for nonreinforced concrete pipe and the required strength for a reinforced concrete pipe.

From Example 5.4: W_d = 2592 lb/ft

nonreinforced

$$\text{Strength } W(\text{3-edge}) = \frac{W_d(\text{SF})}{\text{LF}} = \frac{2952\text{SF}}{\text{LF}}$$

where SF = safety factor = 1.5
LF = load factor for particular bedding class (see Example 5.4)

reinforced Reinforced concrete pipe is designed using the "D load." The D load is the required three-edge strength divided by the pipe diameter. Or

$$\text{Strength } W(D \text{ load}) = \frac{(W_d/D)\text{SF}}{\text{LF}} = \frac{(2592/D)\text{SF}}{\text{LF}}$$

For this material, the strength for each class is based on a 0.01-in crack, not failure. Actual failure load (ultimate) will be approximately 1.5 times the load which causes a 0.01-in crack. Therefore, a safety factor of 1.0 is recommended based on D load or 1.5 based on ultimate load.

Required Strength Based on SF = 1.5 and Ultimate

Bedding class	LF	Three-edge, lb/ft	D-load, (lb/ft)/ft	Nonreinforced	Reinforced
B	1.9	2046	1637	Choose class 1 (2600)	Choose class III (2000)
C	1.5	2592	2074	Choose class 2 (2600)	Choose class IV (3000)
D	1.1	4025	3220	Not available	Choose class V (3750)

NOTE: Above values were calculated and the required classes were selected from Tables 5.5 and 5.6. Also note that a high enough strength for nonreinforced concrete is not available to withstand loads imposed if bedding is only class D.

Example 5.6 (Transmission Pipe) A 24-in diameter transmission line will deliver water at 7000 gal/m and 5.0 ft/s from a reservoir to a treatment plant 10 miles away. The pipe will be buried 5.0 ft deep in a 4.0-ft wide trench in sand carefully compacted or bedded with a coarse granular material up to the spring line. Surge and valve control equipment will allow maximum surge pressures of 50 lb/in². The system will operate at maximum pressure of 150 lb/in². Determine the appropriate prestressed concrete transmission pipe (see Example 5.3).

Prestressed concrete pipe is designed by the cubic parabola method as discussed in Chap. 4. Equation 4.14 is as follows:

$$W = W_o \sqrt[3]{\frac{P_o - p}{P_o}}$$

For lined cylinder: $p = 0.8\, P_o$ (see AWWA C-301)
For embedded cylinder: $p = P_o$ (see AWWA C-301)
From Example 5.3: $W_d = 2112$ lb/ft
Required strength: $W = W_d/\text{LF} = 2112/1.9 = 1111$ lb/ft. $p = 150$ lb/in²

Therefore

$$P_o = 150/0.8 = 187.5 \text{ lb/in}^2 \text{ for lined pipe}$$

Total load $W = W_d + W_L = 2452$ (see Example 5.3)

for lined cylinder pipe $p = 0.8P_o$

$$P_o = 187.5 \text{ lb/in}^2$$

$$W = W_o \sqrt[3]{\frac{P_o - p}{P_o}}$$

$$W_o = \frac{W}{[(P_o - p)/P_o]^{1/3}} = \frac{1111}{[(P_o - 0.8P_o)(P_o)]^{1/3}}$$

or

$$W_o = 1900 \text{ lb/ft}$$

The pipe must be designed or selected for 187.5 lb/in² internal pressure and an external load of 1900 lb/ft each acting independently. The rated strength as determined by the manufacturer includes a safety factor of 1.2. Thus, the transient capacity is considered to be 1.2 times the design capacity for lined cylinder pipe.

$$1.2 \times P_o = 1.2 \times 187.5 = 225 \text{ lb/in}^2$$

$$1.2 \times W_o = 1.2 \times (1900) = 2280 \text{ lb/in}^2$$

Case I (no surge)

$$\text{Max load} = 2280 \sqrt[3]{\frac{225 - 150}{225}} = 1581 \text{ lb/ft}$$

$$\text{Safe live load} = 1581 - 1111 = 470 \text{ lb/ft}$$

Case II (no live load)

$$1111 = 2280 \sqrt[3]{\frac{225 - p}{225}}$$

or

$$p = 225 \left[1 - \left(\frac{111}{2280} \right)^3 \right] = 199 \text{ lb/in}^2$$

Safe surge pressure $= 199 - 150 = 49 \text{ lb/in}^2$

for embedded cylinder pipe   Try

$$W_o = 1900 \text{ lb/ft}$$

$$P_o = 190 \text{ lb/in}^2$$

$$W = W_o \sqrt[3]{\frac{P_o - p}{P_o}} = 1900 \sqrt[3]{\frac{190 - 150}{190}} = 1130 \text{ lb/ft}$$

$$1130 > 1111$$

Thus try is okay. For embedded cylinder pipe, the transient capacity is 1.4 times the design capacity.

$$1.4 \times W_o = 1.4 \times 1900 = 2660 \text{ lb/ft}$$

$$1.4 \times P_o = 1.4 \times 190 = 266 \text{ lb/in}^2$$

Case I (no surge)

$$\text{Max load } w = 2660 \sqrt[3]{\frac{266 - 150}{266}} = 2017 \text{ lb/ft}$$

$$\text{Safe live load} = 2017 - 1111 = 906 \text{ lb/ft}$$

Case II (no live load)

$$P = 266\left[1 - \left(\frac{1111}{2660}\right)^3\right] = 247 \text{ lb/in}^2$$

$$\text{Safe surge pressure} = 247 - 150 = 97 \text{ lb/in}^2$$

These excess capacities are for transient conditions only. The pipe should not be expected to perform with a sustained soil load of 2017 lb/ft nor with a sustained internal pressure of 247 lb/in^2.

Problem 5.1   For the above example (embedded cylinder), try the following combinations of W_o and P_o. For the cases which satisfy design requirements, find the safe live load and safe surge pressure.

W_o	P_o
1800	200
2000	185
1800	190

Flexible Pipes

Thermoplastic pipe materials

There are several types of thermoplastics which are used in the manufacture of pipe. A brief discussion of thermoplastics and design bases are included in Chap. 4. There are four principal thermoplastics used to make pipe:

Polyvinyl chloride (PVC)

Acrylonitrile-butadiene-styrene (ABS)

Polyethylene (PE)

Polybutylene (PB)

Pipes made from other thermoplastics command an extremely small market and are primarily used for specialty applications.

Styrene rubber (SR)

Cellulose-acetate-butyrate (CAB)

Polyvinyl chloride (PVC)   PVC pipe is available for both pressure and gravity applications (Fig. 5.10). For gravity sewer applications, it is available in both solid-wall and profile-wall varieties. Size ranges are as follows:

PVC pressure pipe: ½ to 36 in

PVC solid-wall gravity pipe: 2 to 27 in

PVC profile-wall sewer pipe: 4 to 48 in

The above listed sizes are generally available. However, sizes outside the listed ranges may be available on special order from the manufacturer.

Polyvinyl chloride is manufactured from ethylene and chlorine. Ethylene is extracted from natural gas or crude oil, usually from natural gas. It is also possible to use coal, however that process is much more expensive. Chlorine is manufactured via electrolysis from salt water. Vinyl chloride monomer is produced by oxychlorination (a reaction of ethylene with chlorine). The vinyl chloride monomer (vcm) is polymerized to make polyvinyl chloride (PVC) resin. PVC resin is a white powdery substance with the consistency of table sugar.

This PVC resin is the basic "building block" for PVC pipe. To optimize processability and performance properties, the pipe manufacturer takes PVC resin and compounds it with lubricants, stabilizers, fillers, and pigments. After the mixing takes place at an elevated tem-

Figure 5.10   PVCs high strength to weight ratio is a real advantage. *(Reprinted by courtesy of Uni-Bell PVC Pipe Association.)*

perature, the mixture is allowed to cool to ambient temperature. This PVC compound is fed to a PVC pipe extruder (Fig. 5.11). The extruders are usually of multiscrew design. The PVC compound is worked under high pressure (via extruder screws) and at a controlled elevated temperature so that it is converted into a viscous plastic. A die at the end of the extruder barrel forms the hot viscous plastic into a cylindrical shape. Outside diameter tolerances are maintained by forcing the hot material through a sizing sleeve. After passing through the extruder head and sizing sleeve, the hot pipe is cooled from approximately 400°F as it passes through a spray tank and water bath. The wall thickness and internal diameter dimensions are controlled by balancing the pipe puller speed with the extruder speed. The process is continuous. A cutoff saw which moves with the extruded pipe cuts the pipe in appropriate lengths. The pipe ends are chamfered and the pipe proceeds to a rack where it is positioned for belling (Fig. 5.12).

As explained in Chap. 4, thermoplastics can be heated and reshaped. The pipe belling operation takes advantage of this important property. One end of the pipe is heated and placed in a belling machine where the bell is formed along with a groove for a rubber ring if required. The bell end is then cooled and will maintain its new shape.

The resulting PVC pipe is extremely durable. It is completely inert to water, and to chemicals commonly encountered in sewage and soil

Figure 5.11 PVC pipe extrusion plant. *(Reprinted by courtesy of Uni-Bell PVC Pipe Association.)*

Figure 5.12 PVC pipe belling operation. *(Reprinted by courtesy of Uni-Bell PVC Pipe Association.)*

environments. The surfaces of the pipe are very smooth and resist any buildup of deposited minerals and other solids. It is totally corrosion resistant. It is not attacked by hydrogen sulfide or the resulting sulfuric acid. PVC pipe is not subject to biological degradation. Abrasive resistance is excellent and no special care for cleaning is needed as compared to other pipe products. Dimensional control is excellent and resulting joints are extremely tight. The use of PVC sewer pipe has all but eliminated infiltration and exfiltration and the accompanying tree-root problems.

PVC pipe was first produced and installed on a very limited basis in Germany in the mid 1930s. PVC pipe began to have wide acceptance in the 1960s. Today it commands a large share of the world market, including the market in the United States. It is by far the most widely used plastic pipe. About 90 percent of all plastic pressure-water pipe is PVC and almost 100 percent of plastic sewer pipe is PVC. (Both of these percentages are based on weights shipped.) See Tables 5.8 and 5.9.

PVC gravity sewer pipe. PVC sewer pipe is a flexible pipe and design methods presented in Chap. 3 for flexible pipe are appropriate. Specifically Table 3.5 was developed for any PVC pipe with a pipe stiffness $F/\Delta y = 46$ lb/in^2 and diameter of 4 in through 18 in.

TABLE 5.8 Typical PVC-Pipe Design Properties

Hydrostatic'design basis (HDB)	4000 lb/in^2
Hydrostatic design stress (HDS)	1600 to 2000 lb/in^2
Elastic modulus (pressure formulation)	4,00 lb/in^2
Elastic modulus (sewer formulation)	400,000 sto 550,000 lb/in^2
Tensile stress	7000 lb/in^2
Hazen-Williams coefficient (C)	150
Manning's coefficient (n)	0.009

$$\text{Pipe stiffness PS} = \frac{F}{\Delta y} = \frac{6.7EI}{r^3} = 0.559E\left(\frac{t}{r}\right)^3$$

where

$$r = \text{mean radius} = r_o - \frac{t}{2} = \frac{D_o}{2} - \frac{t}{2} = \frac{D_o - t}{2}$$

Thus,

$$\text{PS} = \frac{F}{\Delta y} = \frac{4.47E}{(DR - 1)^3}$$

where DR = dimension ratio = D_o/t.

For PVC pipes (solid-wall or profile-wall) with diameters larger than 18 in, the manufacturer's recommendations should be obtained and followed. Alternately, Table 3.5 may be used.

Most solid-wall PVC sewer pipes have a DR = 35 and a minimum pipe stiffness of 46 lb/in^2. PVC gravity sewer pipe with pipe stiffnesses in the range of 10 lb/in^2 have been tested and performed adequately when properly installed with a soil density in the pipe zone of at least 85 percent standard proctor density. For any pipe with very low pipe stiffness, extreme care must be taken in preparing and compacting the soil envelope around the pipe. Pipes with less than 10 lb/in^2 pipe stiffness should be used only if a qualified soils engineer is responsible for direction and surveillance of the installation.

Example 5.7 (12-in Gravity Sewer) A 12-in diameter gravity sewer pipe is to be installed in a very deep cut (30 ft). The soil is clay and has been determined to be corrosive and the sewage is septic. Select an appropriate piping material and design the pipe-soil embedment system. The trench width at the top of the pipe may be as much as 4 ft.

1. Calculate soil load (see Chap. 2)

Rigid pipe load

$$W_d = C_d \gamma B_d^2 = 3.3(120)(4)^2$$

$$= 6336 \text{ lb/ft} \quad \text{(see Table 2.2 for } C_d)$$

TABLE 5.9 Standards for PVC Pipe

AWWA C-900	Polyvinyl chloride (PVC) pressure pipe, 4 in through 12 in for water
AWWA C-905	Polyvinyl chloride (PVC) water transmission pipe (nominal diameters 14 to 36 in)
AWWA C-950	Polyvinyl chloride (PVC) water transmission pipe, 14 in through 36 in
ASTM D-2672	Bell-end poly(vinyl chloride) (PVC) pipe
ASTM F-800	Corrugated poly(vinyl chloride) tubing and compatible fittings
ASTM D-3915	Poly(vinyl chloride) (PVC) and related plastic pipe and fitting compounds
ASTM F-679	Poly(vinyl chloride) (PVC) large-diameter plastic gravity sewer pipe and fittings
ASTM F-789	Standard specification for type PS-46 poly(vinyl chloride) (PVC) plastic gravity-flow sewer pipe and fittings
ASTM F-794	Poly(vinyl chloride) (PVC) large diameter ribbed gravity sewer pipe and fittings based on controlled inside diameter
ASTM D-2665	Poly(vinyl chloride) (PVC) plastic drain, waste, and vent pipe and fittings
ASTM D-2466	Poly(vinyl chloride) (PVC) plastic pipe fittings, schedule 40
ASTM D-1785	Poly(vinyl chloride) (PVC) plastic pipe, schedules 40, 80, and 120
ASTM D-2241	Poly(vinyl chloride) (PVC) plastic pipe (SDR-PR)
ASTM D-2740	Poly(vinyl chloride) (PVC) plastic tubing
ASTM D-2729	Poly(vinyl chloride) (PVC) sewer pipe and fittings
ASTM F-599	Poly(vinylidene chloride) (PVDC) plastic-lined ferrous-metal pipe and fittings
ASTM F-656	Primers for use in solvent cement joints of poly(vinyl chloride) (PVC) plastic pipe and fittings
ASTM F-512	Smooth-wall poly(vinyl chloride) (PVC) conduit and fittings for underground installation
ASTM D-3036	Socket-type poly(vinyl chloride) (PVC) plastic line couplings
ASTM D-2467	Socket-type poly(vinyl chloride) (PVC) plastic pipe fittings, schedule 80
ASTM D-3138	Solvent cements for transition joints between acrylonitrile-butadiene-styrene (ABS) poly(vinyl chloride) (PVC) nonpressure piping components
ASTM D-2564	Solvent cements for poly(vinyl chloride) (PVC) plastic pipe and fittings
ASTM F-758	Smooth-wall poly(vinyl chloride) (PVC) plastic underdrain systems for highway, airport, and similar drainage
ASTM F-409	Thermoplastic accessible and replaceable plastic tube and tubular fittings
ASTM D-2464	Threaded poly(vinyl chloride) (PVC) plastic pipe fittings, schedule 80
ASTM F-789	Type PS-46 poly(vinyl chloride) (PVC) plastic gravity-flow sewer pipe and fittings
ASTM D-3034	Type PSM poly(vinyl chloride) (PVC) sewer pipe and fittings
ASTM D-2855	Making solvent cemented joints with poly(vinyl chloride) (PVC) pipe and fittings

Canadian Standards Association	
CSA B 137.0	Definitions, general requirements and methods of testing for thermoplastic pressure piping
CSA B137.3	Rigid poly(vinyl chloride) (PVC) pipe for pressure applications
CSA B181.2	PVC drain waste and vent pipe and pipe fittings
CSA B181.12	Recommended practice for the installation of PVC drain waste and vent pipe and pipe fittings
CSA B182.1	Plastic drain and sewer pipe and pipe fittings
CSA B182.2	Large-diameter, type PSM PVC sewer pipe and fittings
CSA B182.3	Large-diameter, type IPS PVC sewer pipe and fittings
CSA B182.4	Large-diameter, ribbed PVC sewer pipe and fittings

Flexible pipe load

$$\text{Prism load} = \gamma H = 120(30) = 3600 \text{ lb/ft}^2$$

or

$$W = \frac{\gamma H}{B_c} = \frac{\gamma H}{D} = \frac{3600}{1} = 3600 \text{ lb/ft}$$

2. Select piping material. A check will reveal that extra strength clay is not strong enough to withstand the 6336 lb/ft soil load. Also, the highest strength concrete pipe (class 3) is not strong enough. These corrosive conditions would have eliminated concrete and will usually eliminate iron or steel pipe. Use SDR 35 ASTM D-3034 PVC pipe.

3. Design pipe-soil embedment system. For SDR 35 PVC, Table 3.5 may be used for design. The pipe should be installed in a manner such that resulting deflection is less than 7.5 percent. Table 3.5 indicates that class I, class II, or class III soil may be used if compaction is at least 85 percent (see Chap. 3 for definitions of soil classes). Specify class II soil to be used for bedding, haunching, and initial backfill (Fig. 5.13). Pipe-zone soil, to the level of the top of the pipe, must be compacted to at least 85 percent standard proctor density. It is evident from Table 3.5, that the 7.5 percent design deflection will be exceeded if only 80 percent standard proctor density is achieved. It should also be noted that class I soil could have been used without compaction since its natural placement density will be sufficient.

4. Alternate design approach (Spangler's formula), Eq. (3.5):

$$\Delta x = \frac{D_L K W_c r^3}{EI + 0.061 E' r^3} \tag{5.10}$$

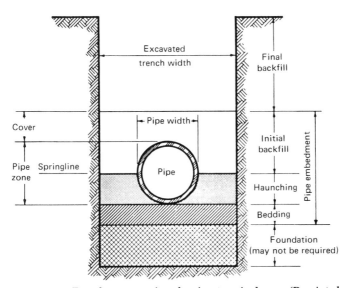

Figure 5.13 Trench cross-section showing terminology. *(Reprinted by courtesy of Uni-Bell PVC Pipe Association.)*

$$\Delta x = \Delta y$$

Assume

$$K = 0.1 \text{ (see Chap. 3 for bedding factors)}$$

$$\frac{W_c}{D} = \gamma H = \text{prism load}$$

$$D_L = 1.0 \text{ when prism load is used}$$

Therefore,

$$\Delta y = \frac{0.1(D\,H)r^3}{EI + 0.061E'r^3}$$

or

$$\frac{\Delta y}{D} = \frac{0.1\gamma H}{EI/r^3 + 0.061E'} \qquad (5.11)$$

$$\text{Pipe stiffness (PS)} = \frac{F}{\Delta y} = \frac{6.7EI}{r^3}$$

or

$$\frac{EI}{r^3} = \frac{PS}{6.7}$$

Equation (5.11) becomes

$$\frac{\Delta y}{D} = \frac{0.1\gamma H}{PS/6.7 + 0.061E'}$$

or

$$\frac{\Delta y}{D} = \frac{0.67\gamma H}{PS + 0.41E'} \qquad (5.12)$$

PS and E' are usually expressed in units of (lb/in^2). If γ is in lb/ft^3 units and H is in ft, γH is lb/ft^2. This must be divided by 144 to convert to lb/in^2.

Assume $\gamma = 120$ lb/ft^3, Eq. (5.12) becomes

$$\frac{\Delta y}{D} = \frac{0.67[(120)(H)/144]}{PS + 0.41E'}$$

$$= \frac{0.56H}{PS + 0.41E'} \qquad (5.13)$$

In the above equation, H is ft of cover. The pipe stiffness PS and soil modulus E' are to be expressed in lb/in^2 units. This equation can be solved for E' as follows:

$$E' = \frac{0.56H/(\Delta y/D) - PS}{0.41} \qquad (5.14)$$

For this example,

$$H = 30 \text{ ft}$$

$$\frac{\Delta y}{D} = 0.075 \text{ (or 7.5 percent)}$$

$$PS = 46 \text{ lb/in}^2$$

Thus,

$$\text{Required } E' = [0.56(30)/0.075 - 46]/0.41 = 434 \text{ lb/in}^2$$

Data in Table 3.4 indicate a soil density of 85 percent is required for finer grain soils with little or no plasticity. Coarse grain soils may be used with little compactive effort required. The two design approaches produce results which agree fairly well. Obviously the use of empirical data from Table 3.5 is the easier method.

Example 5.8 (10-in Gravity Sewer)   A 10-in gravity sewer pipe is to be installed 16 ft deep. The native soil is silty clay and the water table is 10 ft below the surface. Select a PVC pipe and specify the proper installation design if the long-term deflection is not to exceed 7.5 percent.

solution   Select an ASTM D-3034, SDR 35, 10-in sewer pipe. This choice allows the use of Table 3.5 in determining the required embedment soil and soil density. Because of the water table, the trench condition will be wet and the required densities in the pipe zone may not be achievable with native soil. Required compaction must be achieved before high soil loads are imposed and the well points removed. Otherwise the soil will densify with the rising water which may cause excess deflection. However, sufficient backfill must be placed over the pipe (about 3 ft) to prevent flotation of the pipe.

Design I: Use a select clean sand or gravel backfill material (class II Table 3.5) for bedding, haunching, and initial backfill compacted to 85 percent standard proctor density. From Table 3.5, Long-term deflection will be about 3 percent (see Chap. 3 for additional discussion on use of Table 3.5).

Design II: Use a select silty-sandy gravel backfill material (class III, Table 3.5) for bedding, haunching, and initial backfill compacted to 85 percent standard proctor density. From Table 3.5, long-term deflection will be 3.5 percent.

Note   These deflections are substantially lower than the allowed 7.5 long-term deflection. However, because of the wet condition and the relatively deep cover soil, density in the pipe zone must not be less than the density at critical void ratio. This density is often around 90 percent proctor density. For added safety, 90 percent density is recommended. Also design I is preferred to design II because in wet trench conditions the compaction of class III backfill is more difficult.

Example 5.9 (27-in Gravity Sewer)   A 27-in SDR = 35, PS = 46, PVC sewer pipe is to be installed 15 ft deep. The soil is clay except in most areas there is some basalt rock which must be blasted. What type of soil embedment system will be required for this installation?

1. Pipe must not be laid directly on hard pan, bed rock, or any sharp stones with dimensions larger than 1½ in and preferably no stones larger than ¾ in.
2. Excavate at least 6 in below grade and prepare a firm uniform bedding of crushed well-graded stone.

3. Select haunching and initial backfill material: Consider class I, class II, class III, or class IV materials as listed in Table 3.5. A proctor density of 80 percent is sufficient for either class II or class III soils. Class IV soils are often over-looked as pipe embedment materials, but could be used if the trench is not wet and the soil is compacted to 85 percent proctor density. Of course, class I soil will also meet design requirements ($\Delta y/D \leq 7.5$ percent).

4. Spangler's method could also be used but it is not required.

5. Pipe should not be placed directly on sharp rock outcroppings. Also large sharp blasted basalt rock should not be placed directly against the pipe. (A select imported material is recommended.)

Example 5.10 (48-in Ribbed PVC)   A 48-in ribbed PVC pipe is to be installed 20 ft deep. The native soil is fine sand with traces of silt and clay. The pipe stiffness of the ribbed pipe is 10 lb/in². For a special design the owner has requested that this pipe be installed such that the maximum vertical deflection does not exceed 3 percent. Also, to keep costs down, he would like to use the native material for bedding, haunching, and initial backfill. Are these design requirements possible?

solution   Use Spangler's formula.

From Eq. 5.14 of Example 5.7,

$$\text{Required } E' = \frac{0.56H/(\Delta y/D) - \text{PS}}{0.41}$$

$$H = 20 \text{ ft}$$

$$\frac{\Delta y}{D} = 0.03$$

$$\text{PS} = 10 \text{ lb/in}^2$$

$$\text{Required } E' = 886$$

From Table 3.4, required density is 95 percent. This is possible to achieve, but will be difficult to obtain. Owner should be asked to either relax his 3 percent deflection limit or allow a coarser material to be used in pipe zone. Costs associated with compaction may exceed cost of a select material. For a 5 percent deflection limit,

$$E' = \frac{0.56(20)/0.05 - 10}{0.41} = 522 \text{ lb/in}^2$$

For a 7.5 percent deflection limit,

$$E' = \frac{0.56(20)/0.075 - 10}{0.41} = 340 \text{ lb/in}^2$$

The latter can easily be achieved with the native sand used in the pipe zone.

PVC pressure pipe.   PVC pressure pipes are considered to be flexible pipes, and methods presented in Chap. 3 for calculating ring deflection apply. However, most pressure pipes are installed with about 4 ft of cover. Thus the resulting vertical soil pressure is relatively small and consequently ring deflection is usually not a major concern. Only for the lower pressure classes (larger dimension ratios) where the pipe wall is relatively thin and the resulting pipe stiffnesses $(F/\Delta y)$ relatively low is it necessary to consider ring deflection (Table 5.10). As before (see page 174), pipe stiffness is calculated as follows:

$$PS = \frac{F}{\Delta y} = \frac{4.47E}{(DR - 1)^3}$$

The procedure for hydrostatic design is given in Chap. 4. Equation (4.15) is repeated here for convenience.

$$P(D - t) = \sigma \times 2t \qquad (5.15)$$

where P = total internal pressure (static plus surge)
D = outside pipe diameter
t = wall thickness
σ = hydrostatic-design stress

This equation can be rewritten as follows:

$$2\sigma = P(DR - 1) \qquad (5.16)$$

where

$$DR = \frac{D}{t}$$

Equation (5.16) may be solved for DR in terms of hydrostatic-design stress and pressure.

TABLE 5.10 Selected Dimension Ratios (OD/t) and Resulting Pipe Stiffness $(F/\Delta y)$ for PVC Pipes

(OD/t) DR or SDR	Minimum $E = 400{,}000$ lb/in^2	Minimum $E = 500{,}000$ lb/in^2
42	26	32
41	28	35
35	46	57
33.5	52	65
32.5	57	71
28	91	114
26	115	144
25	129	161
21	234	292
18	364	455
17	437	546
14	815	1019
13.5	916	1145

$$DR = \frac{2\sigma}{P} + 1 \qquad (5.17)$$

Example 5.11 (10-in PVC Pressure)   A 10-in PVC pipe is to be used for a transmission pipe in a rural water system. The static pressure will not exceed 150 lb/in². The pipe will be buried in a sandy clay soil with depths between 4 and 5 ft. Select the dimension ratio (DR = OD/t) and design the installation so that the vertical deflection does not exceed 5 percent.

solution

1. The working pressure is 150 lb/in²—no surge pressure needs to be added unless engineer is aware of surge conditions.

2. Assume material is PVC 12454B with a hydrostatic-design basis (HDB) of 4000 lb/in². A safety factor of 2 is required resulting in a hydrostatic-design stress of 2000 lb/in² (see Chap. 4).

3. Use Eq. (5.17) to determine dimension ratio (DR).

$$DR = \frac{OD}{t} = \frac{2\sigma}{P} + 1$$

$$= \frac{2(2000)}{150} + 1 = 27.7$$

Choose next thicker wall from Table 5.10. Use

$$DR = 26$$

$$\frac{F}{\Delta y} = 115 \ lb/in^2$$

4. Determine required pipe-zone material to limit deflection to 5 percent. Data in Table 3.5 indicate that even for loose soil with 5 ft of cover, the maximum deflection will not exceed the 5 percent limit imposed. This table is for pipe with a stiffness of 46 lb/in². For the pipe in this example, the stiffness is 115 lb/in² so it will deflect less. Therefore, no compaction effort is required except for the purpose of limiting surface settlement. The soil placed around the pipe should be free of large stones or frozen lumps.

Example 5.12 (10-in PVC Pressure)   Resolve Example 5.11 for an internal pressure of 100 lb/in² instead of 150 lb/in².

1. Total pressure is 100 lb/in² static plus zero surge pressure

$$p = 100 \ lb/in^2$$

2. Again, PVC 12454B with a hydrostatic-design stress of 2000 lb/in² is selected.

3. Use Eq. (5.12) to determine dimension ratio (DR).

$$DR = \frac{OD}{t} = \frac{2\sigma}{P} + 1$$

$$= \frac{2(2000)}{100} + 1 = 41$$

Therefore, select a PVC pipe where

$$DR = \frac{OD}{t} = 41$$

From Table 5.10,

$$\text{Pipe stiffness (PS)} = \frac{F}{\Delta y} = 28 \text{ lb/in}^2$$

4. Select pipe-zone material and required compaction. Vertical ring deflection is to be less than 5 percent as per Example 5.11. Use Spangler's Equation to determine required E' (see Eq. 5.14)

$$E' = \frac{0.56H/(\Delta y/D) - PS}{0.41}$$

For this example,

$$H = \text{height of cover in feet} = 5 \text{ ft,}$$

$$\frac{\Delta y}{D} = \frac{\text{vertical deflection}}{\text{diameter}} = 0.05$$

$$PS = \text{pipe stiffness} = \frac{F}{\Delta y} = 28 \text{ lb/in}^2$$

Thus,

$$E' = \frac{0.56(5)/0.05 - 28}{0.41} = 68 \text{ lb/in}^2$$

Data in Table 3.4 indicate that a dumped or slightly compacted soil will meet the design criteria. Only uncompacted clays may not meet the specified conditions.

Examples 5.11 and 5.12 indicate that for PVC pressure pipes in medium soil cover, the design of the pipe embedment system is not critical. The primary embedment objective is to protect the pipe from large objects, such as stones, frozen lumps, and objects which could cause impact damage or penetrate the pipe wall.

Example 5.13 (DR 41 PVC)   The DR 41 PVC pipe line operating at 100 lb/in^2 selected in Example 5.12 is to cross a roadway with only 3 ft of cover. Are there special design considerations for this road crossing if a maximum of 2 percent deflection is allowed to protect the road surface?

1. Determine total load.

Total load W_T = prism load

+ live load (See Fig. 2.14 for H20 highway loading.) From Graph, W_T

$$= 950 \text{ lb/ft}^2.$$

2. Use Spangler's equation to calculate required soil modulus E' (see Eq. 5.14)

$$E' = \frac{0.56H/(\Delta y/D) - PS}{0.41}$$

In the above equation, H represents height of cover. For this example, the total load is not just due to soil load but also live load. An effective height H can be calculated as follows:

$$H = \frac{W_T}{120} = 950 \text{ lb/ft}^2/(120 \text{ lb/ft}^3) = 7.9 \text{ ft}$$

Use

$$H = 8 \text{ ft} \quad \frac{\Delta y}{D} = 0.02$$

From previous example:

$$PS = \frac{F}{\Delta y} = 28$$

Thus

$$E' = \frac{0.56(8.0)/0.02 - 28}{0.41} = 478 \text{ lb/in}^2$$

Table 3.4 indicates that a granular material compacted to at least 85 percent proctor density is required. A coarse-grained material with slight compaction will also meet the E' requirement. Experience has shown that for such installations, little or no movement can be tolerated or the road surface will break up. Therefore, a coarse granular material with high compaction is recommended.

Example 5.14a (12-in PVC Pressure)   A 12-in PVC distribution line is to be installed 5 ft deep. The line is to operate at pressures up to 200 lb/in^2. Select the proper dimension ratio and comment on backfill requirements.

1. Calculate design stress. Distribution line, AWWA C-900 applies.

Hydrostatic-design basis (HDB) = 4000 lb/in^2. AWWA safety factor = 2.5.

Hydrostatic-design stress = HDB/SF = 4000/2.5 = 1600 lb/in^2.

2. Determine design pressure.

P = static pressure + surge pressure. AWWA C-900 recommends a 40 lb/in^2 surge pressure for class 200 pipe. Therefore,

$$P = 200 + 40 = 240 \text{ lb/in}^2$$

3. Calculate dimension ratio (DR). Use Eq. 5.12.

$$DR = \frac{2\sigma}{P} + 1$$

$$DR = \frac{2(1600)}{240} + 1 = 14.33$$

Choose $DR = 14$ which has a slightly thicker wall than required.

4. Comment on backfill requirements. Table 5.10 indicates that DR 14 PVC pipe has a pipe stiffness of 815 lb/in^2. This pipe will not require special compaction or soil types when placed with only 5 feet of cover. Compaction may

be necessary to prevent road or surface settlement and to provide soil friction and soil weight to prevent the pipe from floating in saturated soils. Design and construction for thrust restraint will be required at fittings such as elbows and tees (see Chap. 4 for details).

Example 5.14b (Pressure Surge Design)   A water main in a municipal water system with temperatures below 70°F operates with a maximum sustained pressure of 85 lb/in^2. Design engineers predict the maximum instantaneous surge velocity input to be 2 ft/s. For a 12-in diameter pipe, what dimension ratio and corresponding pressure class is required?

1. Try a DR = 18. From AWWA C900, average dimensions are:

Outside diameter OD = 13.200 in

Wall thickness t = 0.733 in

$$ID = OD - 2t = 11.734 \text{ in}$$

2. Calculate wave speed [see Chap. 4 and Eq. (4.9)].

$$a = \frac{4822}{\sqrt{1 + (K/E)(ID/t)}}$$

$$a = \frac{4822}{\sqrt{1 + (313000/400000)(11.734/0.733)}} = 1311 \text{ ft/s}$$

3. Calculate surge pressure P_s.

$$P_s = \left(\frac{a}{g}\right)(V)(0.43) = \left(\frac{1311}{32.2}\right)(2)(0.43) = 35 \text{ lb/in}^2$$

4. Total pressure = working pressure + surge pressure.

$$P = 85 + 35 = 120 \text{ lb/in}^2$$

5. Calculate DR using Eq. (5.12).

$$DR = \frac{2\sigma}{P} + 1 \quad \text{where } \sigma = \frac{HDB}{SF} = \frac{4000}{2.5} = 1600 \text{ lb/in}^2$$

$$= \frac{2(1600)}{120} + 1 = 27.7$$

Select the next available DR which is lower. Use DR = 25 which is AWWA C-900 pressure class 100.

6. Check design with actual dimensions. Use equation in step 2 to recalculate wave velocity.

$$ID = 12.144 \text{ in}$$

$$t = 0.528 \text{ in}$$

$$\text{Wave speed } a = 1106 \text{ ft/s}$$

Use equation in step 3 to recalculate surge pressure.

$$P_s = \left(\frac{1106}{32.2}\right)(2)(0.43) = 29.5 \text{ lb/in}^2$$

Actual surge pressure is lower than used in design calculations. Therefore design is OK.

Example 5.15 (6-in PVC Force Main)   An existing 6-in sewer-force main is to be replaced with a 6-in PVC pressure pipe. The line is known to operate with cyclic pressure surges to a peak pressure of 180 lb/in². The average number of cycles in a 24-h period is 200. The design life of the system is to be a minimum of 50 years. Determine the required dimension ratio (DR).

1. Determine the number of cycles during the life of the system.

Life = 50 years

C' = cycles during life = (50 years)(200 cycles/day)(365 days/year)

= 3.65×10^6 cycles

2. Use Eq. (4.18) to determine required design peak-hoop stress (S').

$$S' = \left[\frac{5.05 \times 10^{21}}{C'} \right]^{0.204}$$

$$= \left[\frac{5.05 \times 10^{21}}{3.65 \times 10^6} \right]^{0.204} = 1230 \text{ lb/in}^2$$

3. Use Eq. (4.19) to calculate required DR.

$$DR = \frac{2S'}{P} + 1 \quad \text{where } P = 180 \text{ lb/in}^2$$

$$= \frac{2(1230)}{180} + 1 = 14.7$$

Select DR = 14 which is AWWA Class 200.

Polyethylene (PE) pipes   Polyethylene used to manufacture pipe is available in several types and grades as per ASTM D-1248. Some grades of polyethylene may crack or craze when subjected to certain levels of stress or when in contact with certain chemicals. This degradation is usually accelerated when high stresses and certain chemicals act simultaneously. This phenomenon is known as environmental stress cracking. Certain grades are highly resistant to stress cracking. Type III, class C, category 5, grade P34 polyethylene is a high density, weather resistant, stress-crack resistant material. (Table 5.11).

TABLE 5.11   Polyethylene Design Properties

Hydrostatic-design basis (HDB)	1250 lb/in²
Hydrostatic-design stress (HDS)	625 lb/in²
Elastic modulus	100,000 lb/in²
Tensile stress (short-time)	3200 lb/in²
Hazen-William coefficient (C)	150
Manning's coefficient (n)	0.009

Polyethylene pipes are available in various sizes and wall configurations for varied applications—some of which are listed in Table 5.12. Other sizes for specific applications may be available from a particular manufacturer. See Table 5.13 for polyethylene standards.

Many of the larger diameter gravity-sewer polyethylene pipes have pipe stiffness (F/y) of 10 lb/in^2 and some even lower than 4 lb/in^2. Extreme care must be taken during installation of these low-stiffness pipes because of the possibility of over-deflection and buckling due to soil load.

Handling factor.   Ring stiffness, the pipe's ability to resist ring deflection, is a function of EI/D^3 (see Chap. 3). Some literature promoting polyethylene pipe gives EI/D^2 as the property which is a measure of the pipe's resistance to deflection. This idea has absolutely no theoretical or experimental bases and if used in the design of a pipe installation could be the direct cause of pipe over-deflection or collapse.

The term (EI/D^2), called "handling stiffness" is sometimes used to rate the ease of handling without damage. The inverse of this factor (D^2/EI) is called the "flexibility factor" and is used by the corrugated-steel pipe industry to rate handling flexibility. These factors arise from a bending strain consideration as follows:

$$\text{Bending strain} = \frac{Mc}{EI} = C_1\frac{(PD)(D/2)(t/2)}{EI}$$

where C_1 = a constant
    P = pressure
    D = diameter
  PD = vertical load
 $D/2$ = moment arm
  $t/2$ = half-wall thickness
    I = wall moment of inertia
    E = modulus of elasticity

One can easily see that D^2/EI is a factor in the above equation. Thus bending strain for a given soil pressure is directly proportional to this

TABLE 5.12   Polyethylene Pipes

Application	Type	Size range, in
Industrial (includes gas)	Solid wall	¾–48
Water (new service)	Solid wall	½–3
Water (insertion)	Solid wall	½–4
Gravity sewer (lining)	Solid wall	4–48
Gravity sewer	Profile (ribbed) wall	18–96

TABLE 5.13 Standards for Polyethylene

ASTM D-3287	Biaxially-oriented polyethylene (PEO) plastic pipe (SDR-PR) based on controlled outside diameter
ASTM D-3261	Butt heat fusion polyethylene (PE) plastic fittings for polyethylene plastic pipe and tubing
ASTM D-405	Corrugated polyethylene tubing and fittings
ASTM D-3197	Insert-type polyethylene fusion fittings for SDR 11.0 polyethylene pipe
ASTM D-2609	Plastic insert fittings for polyethylene (PE) plastic pipe
ASTM D-2104	Polyethylene plastic pipe, schedule 40
ASTM D-2239	Polyethylene plastic pipe, (SIDR-PR) based on controlled inside diameter
ASTM D-3350	Polyethylene plastics pipe and fittings materials
ASTM F-714	Polyethylene plastic pipe (SDR-PR) based on outside diameter
ASTM D-3035	Polyethylene plastic pipe (SDR-PR) based on controlled outside diameter
ASTM D-2447	Polyethylene plastic pipe, schedules 40 and 80 based on outside diameter
ASTM D-2737	Polyethylene plastic tubing
ASTM F-771	Polyethylene thermoplastic high-pressure irrigation pipeline systems
ASTM D-2683	Socket-type polyethylene fittings for outside diameter-controlled polyethylene pipe and tubing
ASTM F-810	Smoothwall polyethylene pipe for use in drainage and waste disposal absorption fields
ASTM D-1248	Polyethylene plastics molding and extrusion materials
AWWA C-901	Polyethylene pressure pipe, tubing and fittings, ½ in through 3 in, for water

factor. The inverse of this factor is a measure of the particular product's ability to resist bending strain. Of course, ring deflection is not a direct function of D^2/EI but is a direct function of D^3/EI. It matters not what causes the deflection—handling, installation, concentrated loads, or soil pressure—the deflection is still a function of D^3/EI, not D^2/EI. Also buckling, whether hydrostatic or due to soil pressure, is a function of D^3/EI. Thus, D^2/EI or EI/D^2 should not be used in design calculations nor should this factor be used to classify a pipe's stiffness characteristics for deflection control.

Example 5.16 (150 lb/in² Polyethylene) Calculate the required dimension ratio (DR) for a polyethylene pressure pipe. The maximum working pressure is 150 lb/in², no surge is anticipated, and the safety factor is to be 2.5.

solution

1. Calculate the hydrostatic design stress:

$$HDS = \frac{HDB}{(\text{safety factor})}$$

HDB = hydrostatic-design bases = 1250 lb/in²

$$\text{HDS} = \frac{1250}{2.5} = 500 \text{ lb/in}^2$$

2. Calculate DR using Eq. (5.17):

$$\text{DR} = \frac{2\sigma}{P} + 1$$

where

$$\sigma = \text{HDS}$$

$$\text{DR} = \frac{2(500)}{150} + 1 = 7.67$$

Select next lower available DR:

$$\text{DR} = 7.0$$

Example 5.17 (6-in Pressure Sewer) It is proposed to use 6-in polyethylene pipe for a pressurized sewer line. The maximum pressure including surge is 50 lb/in^2 and the maximum depth of cover is 20 ft. (*a*) Select the proper wall thickness, (*b*) and what requirements will be necessary concerning pipe-zone soil-type and compaction? (Use a safety factor = 2.0, *OD* = 6.625, and deflection limit = 5 percent.)

solution *a*

1. Hydrostatic-design stress equals (HDB/safety factor)

$$\text{HDS} = \frac{\text{HDB}}{2.0} = \frac{1250}{2} = 625 \text{ lb/in}^2$$

2. Use Eq. (5.17) to calculate dimension ratio (DR):

$$\text{DR} = \frac{2}{P} + 1$$

$$= \frac{2(625)}{50} + 1 = 26$$

$$= 26 = \frac{\text{OD}}{t}$$

$$\text{Thickness} = t = \frac{\text{OD}}{\text{DR}} = \frac{6.625}{26} = 0.25 \text{ in}$$

solution *b*

1. Determine pipe stiffness (*F*/Δ*y*):

$$\frac{F}{\Delta y} = \frac{6.7EI}{r^3}$$

$$E = 100,000 \text{ lb/in}^2$$

$$I = \frac{t^3}{12} = \frac{(0.25)^3}{12}$$

$$r = \text{mean radius} = \frac{OD - t}{2} = 3.19 \text{ in}$$

$$\frac{F}{\Delta y} = \frac{(6.7)(100,000)(0.25/3.19)^3}{12} = 27 \text{ lb/in}^2$$

2. Use Spangler's equation to find required soil modulus (E') (see Eq. 5.14).

$$E' = \frac{0.56H/(\Delta y/D) - PS}{0.41}$$

where PS = pipe stiffness = $F/\Delta y$

$$E' = \frac{0.56(20)/(0.05) - 27}{0.41}$$

$$E' = 480 \text{ lb/in}^2$$

Thus, a granular soil compacted to at least 85 percent proctor density will be required in the pipe zone (see Table 3.4).

Example 5.18 (96-in Storm Sewer)   A 96-in storm sewer is to be installed. The deepest cut will require 14-ft of cover. A profile-wall polyethylene pipe is to be considered. The wall moment of inertia I, of this proposed PE pipe, equals 0.524 in⁴/in. The pipe is to be installed in such a manner that the resulting vertical deflection is less than 5.0 percent. (1) Calculate the pipe stiffness $F/\Delta y$. (2) If selected, how should this particular type of pipe be installed? (3) Comment on the suitability of the proposed pipe for this application.

1. Calculate pipe stiffness $(F/\Delta y)$

$$\frac{F}{\Delta y} = \frac{6.7 \, EI}{r^3}$$

$$E = 100,000 \text{ lb/in}^2$$

$$I = 0.524 \text{ in}^4/\text{in}$$

$$r = \frac{96}{2} = 48 \text{ in}$$

$$\frac{F}{\Delta y} = \frac{(6.7)(100,000)(0.524)}{(48)^3} = 3.17 \text{ lb/in}^2$$

Note: This is a very low value—pipe will be extremely flexible.

2. Design installation: Use Spangler's equation to calculate required E'.* [See Eq. (5.14).]

---

*Note: This equation is derived directly from Spangler's Iowa formula. The Iowa formula is not accurate for very low pipe stiffnesses. Test data at Utah State University indicate that this equation is nonconservative for a pipe stiffness $F/\Delta y = 10$ lb/in², and may not be appropriate for $F/\Delta y = 3.17$. A quick examination of the above equation will show that it cannot hold in the limit as pipe stiffness approaches zero, since it indicates a zero lb/in² pipe it will perform essentially the same as say a 10 lb/in² pipe. Thus, the above equation can be used for pipe stiffness of 10 lb/in² and higher. The error involved is a function of other parameters as well as pipe stiffness. However, the error is within acceptable limits for pipe stiffnesses of 10 lb/in² or greater. Pipes with 3.17 lb/in² pipe stiffness have virtually no inherent strength and stiffness compared with soil. Thus, the pipe in this example should be installed in a well-compacted granular material.

$$E' = \frac{0.56H/(\Delta y/D) - PS}{0.41}$$

$$= \frac{0.56H/(0.05) - 3.17}{0.41}$$

$$= 374 \text{ lb/in}^2$$

It can be determined from Table 3.4 that for $E' = 374$ lb/in^2, a granular material compacted to at least 85 percent proctor density is required.

3. It appears that this particular pipe can be made to work under tightly controlled installation conditions. For pipes with such low stiffnesses, buckling due to soil load is much more likely. This failure mode is discussed in detail in Examples 5.20 and 5.21.

Because pipe stiffnesses below 10 lb/in^2 offer little inherent resistance to deflection, the pipe ring may need to be braced internally while the soil around the pipe is placed and compacted. After the required soil density is obtained, the braces (stills) may be removed. For plastic pipe, bracing may penetrate the pipe wall unless the bracing is carefully designed and positioned. Because of the above concerns, this pipe should be selected only if the above concerns can be addressed. Also a granular material, compacted to at least 85 percent proctor density, should be specified for the pipe zone.

Acrylonitrile-butadiene-styrene (ABS) pipes ABS plastic for pipe manufacture is available in several types and grades as per ASTM D-1788. The physical properties of the various ABS materials vary quite widely as is indicated by Table 5.14. Most ABS pipes, especially pressure pipes, are manufactured from grades with the higher tensile properties.

Solid wall ABS is used widely for drain, waste, and vent piping. It is also used for smaller diameter sanitary sewers. It is used to a very limited extent for smaller diameter pressure piping.

The design methods and procedures are essentially the same as for PVC pipes with the appropriate elastic modulus for calculating pipe stiffness and the appropriate hydrostatic-design stress for pressure-pipe design.

A listing of selected ASTM standards for ABS plastic pipes is given in Table 5.15.

Example 5.19 (8-in ABS) An 8-in solid wall ABS pipe has been selected for a sewer installation. The native soil is clay and the water table is about 8-ft deep.

TABLE 5.14 ABS Design Properties

Hydrostatic-design basis (HDB), lb/in^2	1600–3200
Hydrostatic-design stress (HDS), lb/in^2	800–1600
Elastic modulus, lb/in^2	200,000–400,000
Tensile stress, lb/in^2	2500–7000
Hazen-Williams coefficient (C)	150
Manning's coefficient (n)	0.009

TABLE 5.15   Selected Standards for ABS Plastic Pipe

ASTM D-1788	Rigid acrylonitrile-butadiene-styrene (ABS) plastics
ASTM D-2680	Acrylonitrile-butadiene-styrene composite sewer piping
ASTM D-2661	Acrylonitrile-butadiene-styrene plastic drain, waste and vent pipe, and fittings
ASTM D-628	Acrylonitrile-butadiene-styrene plastic drain, waste and vent pipe having a foam core
ASTM D-2468	Acrylonitrile-butadiene-styrene plastic pipe fittings, schedule 40
ASTM D-1527	Acrylonitrile-butadiene-styrene plastic pipe, schedules 40 and 80
ASTM D-2282	Acrylonitrile-butadiene-styrene plastic pipe (SDR-PR)
ASTM D-2750	Acrylonitrile-butadiene-styrene plastic utilities conduit and fittings
ASTM D-2751	Acrylonitrile-butadiene-styrene sewer pipe and fittings
ASTM D-2469	Socket-type acrylonitrile-butadiene-styrene plastic pipe fittings, schedule 80
ASTM D-2235	Solvent cement for acrylonitrile-butadiene-styrene plastic pipe and fittings
ASTM D-3138	Solvent cements for transition joints between acrylonitrile-butadiene-styrene and poly (vinyl chloride) nonpressure piping components
ASTM D-2465	Threaded acrylonitrile-butadiene-styrene plastic pipe fittings, schedule 80

Most of the line will be installed about 10-ft deep but one section has depths up to 20 ft. The long-term deflection is not to exceed 5 percent. What pipe zone soil and soil density should be specified?

solution   From ASTM D-2751, SDR = 42 and PS = $F/\Delta y$ = 20 lb/in^2. Use Spangler's equation. See Eq. (5.14) of Example 5.7.

$$\text{Required } E' = \frac{0.56H/(\Delta y/D) - PS}{0.41}$$

$$H = 20 \text{ ft} \qquad PS = 20 \text{ lb/in}^2 \qquad \frac{\Delta y}{D} = 0.05$$

$$E' = 498 \text{ lb/in}^2 \approx 500 \text{ lb/in}^2$$

The pipe-zone material should be either a granular material compacted to 90 percent proctor density or a crushed angular stone. Because of the high water table, the crushed stone should be specified since little or no compaction will be required for an angular stone. See Table 3.4 for E' values for various soils.

Other thermoplastic pipes   In addition to the thermoplastic piping materials discussed previously, there are other types of thermoplastic piping materials which are used to a lesser amount. These materials include polybutylene (PB), cellulose acetate butyrate (CAB) and styrene-rubber (SR). A selected list of standards for these materials is given in Tables 5.16, 5.17, and 5.18. The design techniques which are used for thermoplastics such as PVC can also be applied to these thermoplastic materials. The design engineer should obtain necessary design parameters such as the hydrostatic-design stress and pipe stiff-

TABLE 5.16   Selected Standards for Polybutylene

ASTM F-809	Large-diameter polybutylene plastic pipe
ASTM F-809M	Large-diameter polybutylene plastic pipe (metric)
ASTM F-845	Plastic insert fittings for polybutylene (PB) tubing
ASTM D-2662	Polybutylene plastic pipe (SDR-PR)
ASTM D-3000	Polybutylene plastic pipe (SDR-PR) based on outside diameter
ASTM D-2666	Polybutylene plastic tubing
AWWA C-902	Polybutylene pressure pipe, tubing, and fitting, ½ in through 3 in, for water

ness for the particular pipe and material. These parameters may be used in design equations discussed previously.

Example 5.20 (Brittle Behavior)   A strain-sensitive plastic sewer pipe has been installed in an area where expansive soils are known to exist. The pipe deflects as a flexible pipe, has a high pipe stiffness and has a somewhat brittle behavior. A TV inspection made 2 years after installation, indicates vertically elongated pipe with many pipe sections showing longitudinal cracks along the 3 o'clock and 9 o'clock positions. The pipe was installed with a compacted granular material around the pipe, and 10 ft of cover. The expansive soil is to the sides and under the pipe but not over the pipe. The TV photos indicate the pipe to be vertically elongated in the 3 to 8 percent range. Estimate the horizontal-swell pressure exerted by the soil. (Assume $E' = 1000$ lb/in².)

solution   The actual buried pipe may be used as a transducer to obtain a fair estimate of the in situ horizontal swell pressures. This is accomplished by use of the Iowa formula and the actual deflection behavior of the pipe. In short, this formula may be used by providing pipe properties, soil properties, and pipe deflection and then back calculating the pressure necessary to produce that deflection. [see Eqs. (5.11) and (5.12)].

$$\text{Soil modulus } E' = 1000 \text{ lb/in}^2$$

$$\text{Pipe stiffness } \frac{F}{\Delta y} = 200 \text{ lb/in}^2$$

$$\text{Pressure} = (\text{deflection ratio})(10)\left[\frac{\text{pipe stiffness}}{6.7 + 0.061 \text{ (soil modulus)}}\right]$$

or

$$P = \left(\frac{\Delta y}{D}\right)(10)\left[\frac{(F/\Delta y)}{6.7} + 0.061 \, E'\right]$$

The following table indicates probable swell pressures in the range of 23 to 60 lb/in² for deflections of 3 to 8 percent.

Table of Horizontal Swell Pressures for Various Vertical Deflections

Deflection, %	Swell pressure, lb/in²
3	22.8
4	30.4
5	38.0
6	45.5
7	53.1
8	60.7

TABLE 5.17   Selected Standards for Cellulose-Acetate-Butyrate (CAB) Plastic Pipe

ASTM D-2446	Cellulose-acetate-butyrate plastic pipe (SDR-PR) and Tubing
ASTM D-1503	Cellulose-acetate-butyrate plastic pipe, schedule 40
ASTM D-2560	Solvent, cements for cellulose-acetate-butyrate plastic pipe, tubing, and fittings

TABLE 5.18   Selected Standards for Styrene-Rubber (SR) Pipe

ASTM D-3122	Solvent cements for styrene-rubber plastic pipe and fittings
ASTM D-3298	Styrene-rubber plastic drain pipe, perforated
ASTM D-2852	Styrene-rubber plastic drain pipe and fittings

Thermoset plastic pipe

Thermosetting resins give off heat during the curing process (exotherm). Such resins cannot be melted and reformed as thermoplastics can. Epoxy, polyester, and phenolic resins are part of the thermosetting resin family. Pipes made from such resins are usually fiber-reinforced and the fiber is normally "E"-type glass. The glass may be continuous strands or rovings placed in a winding process, or it may be chopped and placed in a centrifugal casting process. Glass fabric and glass mats may also be used.

There are two broad classes of reinforced thermoset pipes: (1) reinforced plastic mortar (RPM) pipe and (2) reinforced thermosetting resin (RTR) pipe. This type has been referred to as fiberglass reinforced plastic (FRP) pipe. The thermoset resin used in either may be filled or unfilled. The filler in the resin is used as a resin extender and will usually influence the chemical and physical properties.

Reinforced thermoset plastic pipe are available in a wide range of sizes. Because of the high tensile strength of the reinforced plastic, a smooth wall pipe may have low pipe stiffness, especially in large diameters. To overcome this, some pipes are made stiffer by molding external ribs which run circumferentially and are spaced along the length. The pipe stiffness is determined with the assumption that the pipe wall and wall stiffeners act integrally as a unit. Such pipes are often designed and manufactured for the specific job with different designs along the installation in response to varying conditions. Table 5.19 gives selected standards for reinforced thermosetting resin pipes. (See Chap. 4 for additional information and design criteria.)

Reinforced thermosetting resin (RTR) pipe   RTR pipes are manufactured from a thermosetting resin and glass fiber reinforcement. The resin may be filled or unfilled. This type of pipe is available in many diam-

TABLE 5.19   Selected Standards for Reinforced Thermosetting Resin Pipe

ASTM D-3517	Reinforced plastic-mortar pressure pipe
ASTM D-3262	Reinforced plastic-mortar sewer pipe
ASTM D-2992	Standard method for obtaining hydrostatic-design basis for reinforced thermosetting resin pipe and fittings
ASTM D-2290	Standard test method for apparent tensile strength of ring or tubular plastics and reinforced plastics by split-disk method
ASTM D-2997	Centrifugally cast reinforced thermosetting resin pipe
ASTM D-2996	Filament-wound reinforced thermosetting resin pipe
ASTM D-2310	Machine-made reinforced thermosetting resin pipe
ASTM D-2517	Reinforced epoxy resin gas pressure pipe and fittings
ASTM D-3840	Reinforced plastic mortar pipe fittings for nonpressure applications
ASTM D-3754	Reinforced plastic mortar sewer and industrial pressure pipe
ASTM D-4160	Reinforced thermosetting resin pipe (RTRP) fittings for nonpressure applications
ASTM D-4163	Reinforced thermosetting resin pressure pipe (RTRP)
ASTM D-4024	Reinforced thermosetting resin (RTR) flanges
ASTM D-4162	Reinforced thermosetting resin sewer and industrial pressure pipe (RTRP)
ASTM D-4184	Reinforced thermosetting resin sewer pipe (RTRP)
ASTM D-1694	Threads for reinforced thermosetting resin pipe
AWWA C-950	Glass-fiber-reinforced thermosetting-resin pipe

eters and for diverse uses for both pressure and nonpressure applications. Liners are available to meet various chemical requirements.

Example 5.21 (84-in Cooling Water)   A fiberglass reinforced polyester resin material has been selected for the pipe to supply cooling water for a large power plant. Selected design parameters are given in Table 5.20. (See AWWA C-950 for design procedures.)

1. Design for deflection

$$\text{Earth load } W_e = (5.5)(110) = 605 \text{ lb/ft}^2$$

TABLE 5.20   Selected Design Parameters

Pipe inside diameter	84 in
Burial depth	5.5 ft (maximum)
Unit weight (soil)	110 lb/ft^3
Live load	300 lb/ft^2
Internal pressure (maximum)	60 lb/in^2
Internal pressure (minimum)	14.7 lb/in^2 vacuum
Water temperature (maximum)	140°F
Hoop modulus (pipe)	3.5 × 10^6 lb/in^2
Bending strain basis	0.0054 in/in
Design strain	0.0036 in/in
Backfill soil	Medium sand at 90% proctor density
Soil modulus E'	Use 650 lb/in^2
Deflection limit	3 percent
Hydrostatic-design basis	10,000 lb/in^2

Live load $W_L = 300$ lb/ft^2

Total load $W = 605 + 300 = 905$ lb/ft$^2 = 6.28$ lb/in^2

Use Spangler's equation to determine required pipe stiffness to control ring deflection. For RTR pipe, a limiting deflection is usually set at some value less than 5 percent. For our problem, the deflection limit has been set at 3 percent. Spangler's equation may be expressed as follows (see Example 5.7):

$$\frac{\Delta y}{D} = \frac{(0.1)(\gamma H)}{PS/6.7 + 0.061\,E'}$$

In this case, H may be replaced by the total load and the above equation will be solved for pipe stiffness (PS).

$$PS = \left[\frac{0.1W}{\Delta y/D} - 0.061E'\right]6.7$$

For $W = 6.28$ lb/in^2, $\Delta y/D = 0.03$, and $E' = 650$ lb/in^2, the pipe stiffness PS is found to be negative therefore deflection does not control design. This conclusion is based on the assumption that the pipe will be installed properly with a resulting E' equal to 650 lb/in^2.

2. Assume that the pipe may not be installed as per design specifications. What is the minimum soil modulus E' that can be accepted and still meet the 3 percent deflection limit (assume pipe stiffness PS = 10 lb/in^2).

Use Spangler's equation to solve for E'.

$$E' = \left[\frac{(0.1W)}{(\Delta y/D)} - \frac{(PS)}{6.7}\right]\left(\frac{1}{0.061}\right)$$

$$= \left[\frac{(0.1)(6.28)}{0.03} - \frac{10}{6.7}\right]\left(\frac{1}{0.061}\right)$$

$$= 319 \text{ lb/in}^2$$

3. Design for buckling (see AWWA C-950). The buckling equation given in AWWA C-950 is as follows:

$$q_a = \frac{[32R_WB'E'(EI/D^3)]^{1/2}}{SF}$$

or

$$q_{cr} = [32R_WB'E'(EI/D^3)]^{1/2}$$

where q_a = allowable buckling pressure
   SF = safety factor or design factor usually taken as 2.5 or greater
   R_W = water buoyancy factor, 1.0 for our problem
   B' = empirical coefficient of elastic support (dimensionless)
      = $1/(1 + 4e^{-0.065H})$ = 0.26 for our problem
   H = burial depth to top of pipe, ft

Buckling pressure $q_a = 14.7$ lb/in^2 vacuum + 6.28 lb/in^2 soil pressure

$$= 20.98 \approx 21 \text{ lb/in}^2$$

Use the AWWA equation to solve for EI/D^3.

$$\frac{EI}{D^3} = \frac{q_a{}^2(\text{SF})^2}{(32R_W B'E')}$$

$$= \frac{(21)^2(2.5)^2}{(32)(1)(0.26)(650)}$$

$$= 0.51 \text{ lb/in}^2$$

$$\text{Pipe stiffness } PS = 6.7\left(\frac{EI}{r^3}\right) = 6.7\left(\frac{EI}{D^3}\right)(8)$$

$$= 53.6\left(\frac{EI}{D^3}\right)$$

Therefore, the required pipe stiffness is

$$PS = (53.6)(0.51) = 27.34 \text{ lb/in}^2$$

$$q_{cr} = [32(1.0)(0.26)(650)(0.51)]^{1/2} = 52.5 \text{ lb/in}^2$$

The thickness required for a straight wall pipe may be determined using above stiffness as follows:

$$PS = 6.7\left(\frac{EI}{r^3}\right) = 53.6\left(\frac{EI}{D^3}\right)$$

or

$$I = \frac{(\text{PS})D^3}{53.6E}$$

but

$$I = \frac{t^3}{12}$$

then

$$t^3 = \frac{12(\text{PS})D^3}{53.6E}$$

or

$$t = 0.61D(\text{PS})^{1/3}E^{-1/3}$$

$$= 0.61D(\text{PS})^{1/3}(3.5 \times 10^6)^{-1/3}$$

$$= 1.02 \text{ in}$$

4. Check pressure design. Internal pressure including surge is given to be 60 lb/in². A quick check on stress due to internal pressure reveals a low value.

$$\sigma = \frac{PD}{2t} = \frac{(60)84}{2(1.02)} = 2471 \text{ lb/in}^2$$

$$2471 \lll 10,000$$

Thus, stress due to internal pressure acting alone is not a critical factor.

5. Check strain due to ring deflection. The bending strain caused by the 3 percent design-ring deflection is calculated using Eq. (3.19).

$$\epsilon_b = 6\left(\frac{t}{D}\right)\left(\frac{\Delta y}{D}\right)$$

$$= 6\left(\frac{1.02}{84}\right)(0.03)$$

$$= 0.00219$$

The above strain is less than the 0.0036 design strain.

6. Calculate strain due to combined loading. (See Chap. 4—fiber reinforced plastic and AWWA C-950.) Two equations are given in AWWA C-950 for calculating strain due to the simultaneous action of ring bending and internal pressure. The so-called "Molin equation" is to be used for low pressures and another equation based on Spangler's Iowa formula is to be used for higher pressures. The maximum strain is the lower of the two calculated values. For our problem, the internal pressure is quite small, therefore, the equation attributed to Molin applies.

$$\text{Combined strain } \epsilon_c = \frac{PD}{2Et} + 6\left(\frac{\Delta y}{D}\right)\left(\frac{t}{d}\right)$$

This equation is just the simple addition of the strain due to internal pressure with the strain due to ring bending—a simple concept of elementary mechanics of materials.

For the problem at hand,

$$\epsilon_c = \frac{(60)(84)}{[2(3.5 \times 10^6)(1.02)]} + 6(0.03)\left(\frac{1.02}{84}\right)$$

$$= 0.706 \times 10^{-3} + 2.19 \times 10^{-3} = 2.90 \times 10^{-3}$$

or

$$\epsilon_c = 0.00290$$

This is less than the design strain of 0.0036. Thus, combined strain is all right.

Example 5:22 (84-in Ribbed)   In Example 5.21, it was determined that the wall thickness should be 1.02 in. This was the thickness required to produce a pipe stiffness $PS = 27.34$ lb/in^2 which was required for buckling design. Suppose a ribbed pipe is to be used instead of the solid-wall pipe designed in the previous example. The ribbed pipe is to have ribs spaced on 78-in centers and the wall thickness between ribs is to be 0.6 in. The ribs will be constructed to act in an integral manner with the wall such that the pipe stiffness, PS, is equal to 27.34 lb/in^2 as required previously.

Carry out necessary calculations to determine if the ribbed pipe will perform adequately.

1. Check pressure design (see Example 5.21).

$$\sigma = \frac{PD}{2t} = \frac{(60)(84)}{2(0.6)} = 4200 \text{ lb/in}^2$$

Since the hydrostatic design basis (HDB) = 10,000 lb/in^2, the safety factor is 10,000/4200 = 2.38.

2. Check bending strain (see Example 5.20).
   a. Find strain in wall at a point away from rib.

$$\epsilon_b = 6\left(\frac{t}{D}\right)\left(\frac{\Delta y}{D}\right)$$

$$= 6\left(\frac{0.6}{84}\right)(0.03) = 1.29 \times 10^{-3} \text{ in/in}$$

   b. Find strain in wall at point near rib. Assume rib thickness from inside wall to outside of rib is 2.10 in and also assume the distance from the inside wall to the centroid of the wall section is $X_c = 0.68$ in.
   Since the wall thickness is 0.60 inch, the centroid is 0.08 in outside of wall.

$$\epsilon_b = 6\left(\frac{t}{D}\right)\left(\frac{\Delta y}{D}\right) = 12\left(\frac{t}{2D}\right)\left(\frac{\Delta y}{D}\right)$$

where $t/2$ may be replaced by 0.68. Thus,

$$\epsilon_b = 12\left(\frac{0.68}{84}\right)(0.03) = 2.91 \times 10^{-3} \text{ in/in}$$

Wall bending strain is within design limits.

3. Check combined strain (see Example 5.21). For near rib:

$$\epsilon_c = \frac{PD}{2Et} + 6\left(\frac{t}{D}\right)\left(\frac{\Delta y}{D}\right)$$

$$= \frac{PD}{2Et} + 12\left(\frac{t}{2D}\right)\left(\frac{\Delta y}{D}\right)$$

where $t/2$ can be replaced by 0.68 in (see Example 5.20).

$$\epsilon_c = \frac{(60)(84)}{2(3.5 \times 10^6)(0.6)} + 12\left(\frac{0.68}{84}\right)(0.03)$$

$$= 1.20 \times 10^{-3} + 2.91 \times 10^{-3}$$

$$= 4.11 \times 10^{-3} \text{ in/in}$$

This strain exceeds the design strain of 3.6×10^{-3}. However, the design strain included a safety factor and the pressure used included a surge pressure. Also, the effective thickness near the rib is larger than the 0.6 used in the calculation. In any case, the limiting long-term strain of 5.4×10^{-3} in/in has not been exceeded so combined strain is all right.
   In wall away from rib:

$$\epsilon_c = \frac{PD}{2Et} + 6\left(\frac{t}{D}\right)\left(\frac{\Delta y}{D}\right)$$

$$= \frac{(60)(84)}{[2(3.5 \times 10^6)(0.6)]} + 6\left(\frac{0.6}{84}\right)(0.03)$$

$$= 1.20 \times 10^{-3} + 1.29 \times 10^{-3}$$

$$= 2.49 \times 10^{-3}$$

4. Check buckling. The ribbed pipe in this example has the same pipe stiffness as the solid wall pipe of Example 5.21. Therefore, general buckling will not occur and a design check should be made for localized buckling. Texts, dealing with advanced mechanics of materials or theory of elasticity, usually have solutions for localized buckling of tubes with ring stiffeners. The book, *Theory of Elastic Stability*, by Timoshenko and Gere, gives such a solution in graphical form on page 480 (see Fig. 5.14). These solutions are for tubes subjected to hydrostatic pressure and *not* constrained by soil. The surrounding soil effectively stiffens the pipe. Thus, a pipe in soil will take a larger buck-

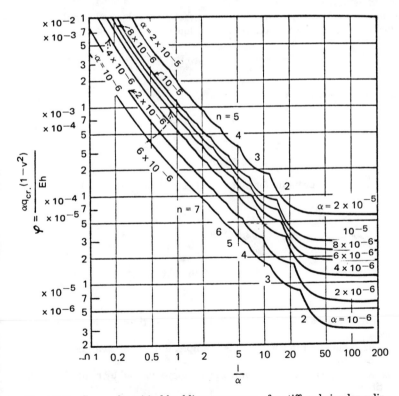

Figure 5.14 Curves for critical buckling pressure q_{cr} for stiffened circular cylinders subjected to a uniform radial pressure. l = rib spacing = 78 in, a = pipe radius = 42 in, ν = Poisson's ratio = 0.3, n = pipe thickness = 0.6 in., $\alpha = t^2/12r^2$ = $(0.6)/[(12)(42)^2]$, E = elastic modulus (3.5×10^6 lb/ft.^{2}), q_{cr} = buckling pressure.) (*Reprinted by permission from Timoshenko and Gere,*[14] *Elastic Stability, McGraw-Hill, New York, 1961.*)

ling load as compared to a pipe subjected to hydrostatic pressure. Therefore, the hydrostatic solutions are conservative.

From Fig. 5.14, we can determine the following:

$$\alpha = \frac{t^2}{12r^2} = \frac{(0.6)}{(12)(42)^2} = 1.7 \times 10^{-5}$$

From the curves, $\psi = 0.9 \times 10^{-4}$

$$q_{cr} = \frac{\psi\, Eh}{a(1 - \nu^2)}$$

$$= \frac{(9 \times 10^{-4})(3.5 \times 10^6)(0.6)}{(42)(1 - 0.9)}$$

$$= 49.5 \text{ lb/in}^2$$

Again, this is the buckling pressure for a pipe subjected to hydrostatic pressure without soil support. The actual buckling pressure will be larger and can be approximated as follows:

The general buckling pressure for a long tube (pipe) subjected to only hydrostatic loading is given by the following equation:

$$q_{cr} = \frac{3EI}{r^3(1 - \nu^2)} \qquad \text{(see Eq. 3.13)}$$

For the pipe in our example,

$$q_{cr} = \frac{3(EI/r^3)}{1 - \nu^2} = \frac{3(4.08)}{0.91}$$

$$\doteq 13.5 \text{ lb/in}^2$$

The general buckling pressure for pipe in soil as calculated in part 3 of Example 5.21 was

$$q_{cr} = \left[32\, R_w B' E' \left(\frac{EI}{D^3} \right) \right]^{1/2}$$

$$= [32(1.0)(0.26)(6.50)(0.51)]^{1/2}$$

$$= 52.5 \text{ lb/in}^2$$

Note that this pressure is $52.5/13.5 = 3.9$ times greater than for the pipe with no soil support.

For localized buckling in soil, for this example, a factor of 2 can be used conservatively.

Thus, the localized buckling pressure can be approximated by multiplying the hydrostatic value by 2.

$$q_{cr} = (49.5)(2) = 99 \text{ lb/in}^2$$

The applied pressure is 21 lb/in^2 (see Example 5.21).

The pipe, in this example, will not experience localized buckling. General buckling will occur at a lower pressure than localized buckling. In fact, localized buckling will not occur even without soil support. A note of caution: The above analyses assume a fairly uniform pressure. Nonuniform pressures or high pressure concentration will substantially lower the critical buckling pressures and may lead to localized buckling. Extreme hard spots such as large rocks or other hard debris next to the pipe can cause such pressure concentrations. These can be avoided by proper construction practices. Nonuniform pressures will occur, however, and are difficult to predict quantitatively. To compensate for such unknowns, safety factors are required.

Steel pipe

Steel pipe is used in many diverse applications. It is available in various sizes, shapes, and wall configurations. For pressure application, the cross section is circular. However, for gravity flow, steel pipes can have cross sections which are vertical elongated ellipses, arch-shaped for low head room, so called "long-span arched sections," and other shapes.

For the most part, steel pipes used for gravity applications have a corrugated wall. The corrugated shape produces a larger moment of inertia which results in a larger pipe stiffness. Such pipes are usually galvanized for corrosion protection but are also available as aluminized steel. Common coatings and linings available include bitumen-type materials, portland cement-type materials, and polymers. In certain applications, the lining may be applied after installation. The linings and coatings are usually ignored in strength and stiffness calculations.

Design information for corrugated-steel products is available in the *Handbook of Steel Drainage and Highway Construction Products* which is published by the American Iron and Steel Institute. Also, many manufacturers publish design information for their products. Such information should be secured and considered by the designer. For corrugated-steel pipes with circular sections, standard analysis and design procedures which have been discussed in this book apply and may be used by the design engineer.

Example 5.23 (48-in Corrugated Steel)   A 48-inch diameter (3" by 1") corrugated-steel pipe is to be placed in an embankment with 60 ft of soil cover. The soil in the pipe zone is to be coarse sand with some fines and is to be compacted to 90 percent proctor density.

1. What thickness is required so that the pipe deflection does not exceed five percent?

solution   Use Spangler's equation.

$$\frac{\Delta y}{D} = \frac{(0.1)(\gamma H)}{EI/r^3 + 0.061E'}$$

$$H = 60 \text{ ft}$$

Let

$$D = 48 \text{ in}$$

$$\gamma = 120 \text{ lb/ft}^3$$

$$E = 30 \times 10^6 \text{ lb/in}^2$$

$$E' = 1000 \text{ lb/in (from Table 3.4)}$$

Solve for EI/r^3.

$$\frac{EI}{r^3} = \frac{(0.1)(\gamma H)}{(\Delta y/D)} - 0.061E'$$

$$= \frac{(0.1)(120)(60)(1/144)}{0.05} - 0.061(1000)$$

or

$$= 100 - 61 = 39$$

$$I = \frac{39r^3}{E} = \frac{39(24)^3}{30 \times 10^6}$$

$$= 0.018 \text{ in}^4/\text{in}$$

$$= 0.22 \text{ in}^4/\text{ft}$$

From Table 5.21, the uncoated thickness should be 0.1345 in.

2. Assume the yield stress (σ_y) for the steel is 33,000 lb/in². What wall area is required for ring compression design with a safety factor of 2?

solution

$$\text{Design compression stress } f_c = \frac{\sigma_y}{2} = 16,500 \text{ lb/in}^2$$

$$\text{Vertical soil pressure } P_V = (120)(60) = 7,200 \text{ lb/ft}^2$$

or

$$P_V = \frac{7200}{144} = 50 \text{ lb/in}^2$$

$$f_c = \frac{PDL}{2A} = \frac{PD}{(2A/L)}$$

Solve for A/L.

$$\frac{A}{L} = \frac{PD}{2f_c} = \frac{(50)(48)}{2(16,500)} = 0.073 \text{ in}^2/\text{in}$$

or

$$\frac{A}{L} = 0.073 \text{ in}^2/\text{in}(12 \text{ in/ft}) = 0.88 \text{ in}^2/\text{ft}$$

From Table 5.21, the uncoated thickness is 0.0598 in. Thus, the deflection design controls and the thickness found in 1. is the required thickness.

Steel pressure pipes are used in many varied and diverse applications in industrial, agricultural, and municipal markets. The discussion here will

TABLE 5.21 Sectional Properties of Corrugated Steel Sheets

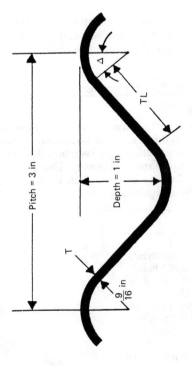

Specified thickness in	Uncoated thickness I, in	Area of section A, in²/ft	Tangent length TL, in	Tangent angle Δ, degrees	Moment of inertia† I, in⁴/ft	Section modulus† S, in³/ft	Radius of gyration r, in	Developed width‡ factor
0.040*	0.0359	0.534	0.963	44.19	0.0618	0.1194	0.3403	1.239
0.052	0.0478	0.711	0.951	44.39	0.0827	0.1578	0.3410	1.240
0.064	0.0598	0.890	0.938	44.60	0.1039	0.1961	0.3417	1.240
0.079	0.0747	1.113	0.922	44.87	0.1306	0.2431	0.3427	1.241
0.109	0.1046	1.560	0.889	45.42	0.1855	0.3358	0.3448	1.243
0.138	0.1345	2.008	0.855	46.02	0.2421	0.4269	0.3472	1.244
0.168	0.1644	2.458	0.819	46.65	0.3010	0.5170	0.3499	1.246

*Thickness not commonly available. Information only.
†Per foot of projection about the neutral axis. To obtain A, I, or S per inch of width, divide by 12.
‡Developed width factor measures the increase in profile length due to corrugating. Dimensions are subject to manufacturing tolerances.

TABLE 5.22   Selected Standards for Steel Pressure Pipes in Water Service

AWWA C-200	Steel water pipe 6 in and larger
AWWA C-203	Coal-tar protective coatings and linings for steel water pipelines— enamel and tape-hot applied
AWWA C-205	Cement-mortar protective lining and coating for steel water pipe— 4 in and larger—shop applied
AWWA C-206	Field welding of steel water pipe
AWWA C-207	Steel pipe flanges for waterworks service—sizes 4 in through 144 in
AWWA C-208	Dimensions for fabricated steel water pipe fittings
AWWA C-209	Cold-applied tape coatings for special sections, connections, and fittings for steel water pipelines
AWWA C-210	Coal-tar epoxy coating system for the interior and exterior of steel water pipe
AWWA C-213	Fusion-bonded epoxy coating for the interior and exterior of steel water pipelines
AWWA C-214	Tape coating systems for the exterior of steel water pipelines
AWWA M-11	Steel pipe design and installation

be limited to steel pipe used primarily in the municipal water market (see Table 5.22). However, principles used are applicable to all steel pressure pipe.

Example 5.24 (108-in Transmission)   A 108-in diameter water-transmission line is to be installed. Steel has been selected as the piping material. The joint is to be a bell- and spigot-type joint welded both inside and out as shown:

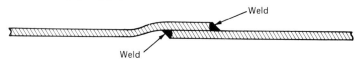

Weld

Weld

The wall thickness is to be 0.5 in. Because of the large diameter, the pipe will be very flexible and will be braced with internal bracing (stills) when manufactured. These stills will remain in the pipe sections until they have been installed and pipe zone soil has been placed and compacted to the specified density. The stills will be removed after backfilling is complete. The pipe line will then be lined with a portland cement-type mortar before the line is placed in service.

Design Parameters	
Wall thickness	0.5 in
Yield stress	36,000 lb/in^2
Ultimate strength	60,000 lb/in^2
Modulus	29 × 10^6 lb/in^2
Poisson's ratio	0.3
Thermal coefficient of expansion	6.5 × 10^{-6}(1/°F)
Ductile-brittle transition temperature	(70°F)
Surge pressure allowance	40 lb/in^2
Cover depth	6 ft
Pipe-zone soil	Crushed stone
Pipe-zone density	90% standard proctor
Water temperature	34°F

Evaluate the proposed steel pipe for this application. Are there any special precautions which should be taken or special construction methods which should be followed.

1. Check pipe stiffness and evaluate possible ring deflection.

$$PS = \frac{F}{\Delta y} = \frac{6.7EI}{r^3}$$

$$= \frac{6.7(29 \times 10^6)(0.5)^3}{(12)(54)^3}$$

$$= 12.85 \ lb/in^2$$

This pipe is quite flexible. However, the pipe is going to be held in the undeflected state until pipe-zone soil is compacted and the overburden is placed. The resulting deflection after the stills are removed will be quite low.

2. Check pressure design.
   a. Find hoop stress for design pressure plus surge.

$$\sigma_h = \frac{PD}{2t} = \frac{(120 + 40)(108)}{2(0.5)}$$

$$= 17,280 \ lb/in^2$$

   b. Find hoop stress for design pressure only.

$$\sigma_h = \frac{PD}{2t} = \frac{(120)(108)}{2(0.5)}$$
$$= 12,960 \ lb/in^2$$

The yield stress is 36,000 lb/in^2. The safety factor is greater than 2, therefore, pressure design is all right.

3. Consider longitudinal stresses. AWWA C-206 indicates that temperature considerations should be made in design. AWWA C-206 and AWWA M-11 suggest the use of either closure welds or expansion joints to alleviate stresses due to temperature change.

   Longitudinal stresses will also be produced by the Poisson effect. Temperature stresses and Poisson stresses along with bending stresses due to nonparallel loading in the bell-spigot connection, may be large enough to cause failure.

   Assume the pipe is placed and tack welded during the day. It is July and August and the pipe temperature during tack welding is between 80 and 130°F. The tack welds hold firm and the welding process is completed by a welding crew who are following behind the pipe laying crew. No closure welds or expansion joints are being used. After the line is completed, it is put in service with water at 120 lb/in^2 and 34°F. (See Chap. 4—longitudinal stresses and steel pipe sections.)

   a. Find the longitudinal stress due to the Poisson effect.

$$\sigma_p = \nu\sigma_h \quad \text{but} \quad \sigma_h = 12,960 \ lb/in^2$$

$$= (0.3)(12,960) = 3,888 \ lb/in^2$$

b. Find the longitudinal stress due to temperature change.

$$\sigma_T = E\alpha(\Delta T)$$

$$= (29 \times 10^6)(6.5 \times 10^{-6})(\Delta T)$$

$$= (188.5)(\Delta T)$$

Assume a $\Delta T = 70°F$

$$\sigma_T = 13{,}195 \text{ lb/in}^2$$

c. What is the total longitudinal stress?

$$\sigma_L = \sigma(\text{Poisson}) + \sigma(\text{temperature})$$

$$= 3888 + 13{,}195 = 17{,}083 \text{ lb/in}^2$$

d. The nonparallel loading in the bell-spigot will produce a bending moment and will effectively magnify the stress found in (*c*) above. What is that magnification factor?

$$\text{Bending stress} = \sigma_B = \frac{MC}{I}$$

where M = moment = $(\sigma_L)(A)(t)$ = $\sigma_L(bt)(t)$
t = thickness
A = area = bt
$C = t/2$
$I = bt^3/12$

Therefore

$$\sigma_B = \frac{(\sigma_L)(bt)(t)(t/2)}{bt^3/12}$$

$$= 6\sigma_L$$

Then, the bending stress is six times the longitudinal stress. However, the maximum stress is the sum of the bending stress plus the longitudinal stress.

$$\sigma_{\max} = \sigma_B + \sigma_L = 7\sigma_L$$

The magnification factor is 7. Therefore, $\sigma_{\max}$ = (7)(17,083) = 119,581 lb/in².

The pipe will fail before this stress is reached. In fact, it did. This pipeline was actually designed and constructed as described in this example. The designer failed to consider longitudinal stresses and did not allow for closure or expansion joints. There were three separate failures caused by longitudinal stresses. Each time a repair was made the line was returned to service. After the third failure, a general repair was ordered. Every other joint was cut to relieve the built-in stresses. As the joints were cut, there were snap back openings of as much as 1 in. The temperature of the pipe during the repair was 55°F, 21 degrees higher than the service temperature, so there will still be some stress at 34°F. Had the steel been more ductile it may have been able to relieve itself by simply stretching. For the steel selected, the ductile-brittle transition temperature was 70°F. Therefore, the steel behaved in a brittle manner and failed.

Ductile-iron pipe

Ductile-iron pipe has essentially replaced gray cast-iron pipe. Ductile iron is, as its name implies, more ductile than gray cast iron but still retains somewhat brittle properties. It is very popular among public works people who repair and maintain water systems. Many of them perceive this pipe as able to withstand abuse during handling and repair operations.

Corrosion rate for ductile iron is essentially the same as for gray cast iron. However, since the wall is usually thinner, corrosion is more critical. Design procedures call for a corrosion allowance, called a "service factor." When installed in highly corrosive soil, steps should be taken to protect it. Ductile-iron pipe is usually lined with a cement-mortar lining. This lining improves the hydraulic efficiency and also provides some corrosion protection. Other linings and coatings are available. See Table 5.23.

Example 5.25 (30-in DI Pipe)   Calculate the thickness for 30-in ductile-iron (DI) pipe laid on a flat-bottom trench with backfill tamped to centerline of pipe, laying condition type 2 (Fig. 5.15), under 10 ft of cover for a working pressure of 200 lb/in². (See ductile iron section in Chap. 4 for design procedure for pressure pipe. Also, see AWWA C-150. Certain tables from AWWA C-150 have been reproduced here for the reader's convenience. This example is taken from AWWA C-150).

1. Design for trench load
   a. Earth load, Table 5.24, P_e = 8.3 lb/in². May be obtained from Fig. 2.14. Truck load, Table 5.24, P_t = 0.7 lb/in². Trench load, $P_v = P_e + P_t$ = 9.0 lb/in²
   b. Select Table 5.29 for diameter-thickness ratios for laying condition type 2.
   c. Entering P_v of 9.0 lb/in² in Table 5.29, the bending stress design requires D/t of 128. From Table 5.28, diameter D of 30-in pipe, OD is 32.00 in. Net thickness t for bending stress is

$$t = \frac{D}{(D/t)} = \frac{32.00}{128} = 0.25 \text{ in}$$

TABLE 5.23   Selected Standards for Ductile Iron Pipe

AWWA C-104	Cement mortar lining for ductile iron
AWWA C-105	Polyethylene encasement for ductile iron
AWWA C-110	Ductile iron and gray iron fittings
AWWA C-111	Rubber-gasket joints for ductile iron
AWWA C-115	Flanged ductile iron
AWWA C-150	Thickness design of ductile iron pipe
AWWA C-151	Ductile iron pipe in metal and sandlined molds
ASTM E-8	Materials properties test
ASTM A-539	Physical properties

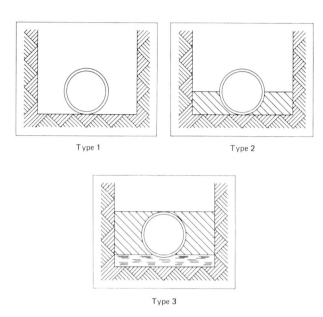

Type 1   Type 2

Type 3

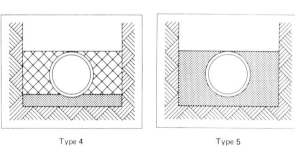

Type 4   Type 5

Figure 5.15 Standard pipe laying conditions. *(Reprinted, by permission, from ANSI/AWWA C150/A121.50-81 (R-86), American Water Works Association, 1986.)*

d. Also, from Table 5.29, the deflection design requires D/t_1 of 108. Minimum thickness t_1 for deflection design:

$$t_1 = \frac{D}{D/t_1} = \frac{32.00}{108} = 0.30 \text{ in}$$

Deduct service allowance – 0.08 in

Net thickness t for deflection control = 0.22 in

e. The larger net thickness is 0.25 in, obtained by the design for bending stress.

2. Design for internal pressure:

$P_i = 2.0$ (working pressure + 100 lb/in² surge allowance)

TABLE 5.24

Depth of cover, ft	P_e	18-in pipe		20-in pipe		24-in pipe		30-in pipe	
		P_t	P_v	P_t	P_v	P_t	P_v	P_t	P_v
2.5	2.1	7.8	9.9	7.5	9.6	7.1	9.2	6.7	8.8
3	2.5	5.9	8.4	5.7	8.2	5.4	7.9	5.2	7.7
4	3.3	3.9	7.2	3.9	7.2	3.6	6.9	3.5	6.8
5	4.2	2.6	6.8	2.6	6.8	2.4	6.6	2.4	6.6
6	5.0	1.9	6.9	1.9	6.9	1.7	6.7	1.7	6.7
7	5.8	1.4	7.2	1.4	7.2	1.3	7.1	1.3	7.1
8	6.7	1.2	7.9	1.1	7.8	1.1	7.8	1.1	7.8
9	7.5	1.0	8.5	0.9	8.4	0.9	8.4	0.9	8.4
10	8.3	0.8	9.1	0.7	9.0	0.7	9.0	0.7	9.0
12	10.0	0.5	10.5	0.5	10.5	0.5	10.5	0.5	10.5
14	11.7	0.4	12.1	0.4	12.1	0.4	12.1	0.4	12.1
16	13.3	0.3	13.6	0.3	13.6	0.3	13.6	0.3	13.6
20	16.7	0.2	16.9	0.2	16.9	0.2	16.9	0.2	16.9
24	20.0	0.1	20.1	0.1	20.1	0.1	20.1	0.1	20.1
28	23.3	0.1	23.4	0.1	23.4	0.1	23.4	0.1	23.4
32	26.7	0.1	26.8	0.1	26.8	0.1	26.8	0.1	26.8

Depth of cover, ft	P_e	36-in pipe		42-in pipe		48-in pipe		54-in pipe	
		P_t	P_v	P_t	P_v	P_t	P_v	P_t	P_v
2.5	2.1	6.2	8.3	5.8	7.9	5.4	7.5	5.0	7.1
3	2.5	4.9	7.4	4.6	7.1	4.4	6.9	4.1	6.6
4	3.3	3.4	6.7	3.3	6.6	3.1	6.4	3.0	6.3
5	4.2	2.3	6.5	2.3	6.5	2.2	6.4	2.1	6.3
6	5.0	1.7	6.7	1.7	6.7	1.6	6.6	1.6	6.6
7	5.8	1.3	7.1	1.3	7.1	1.2	7.0	1.2	7.0
8	6.7	1.1	7.8	1.0	7.7	1.0	7.7	1.0	7.7
9	7.5	0.8	8.3	0.8	8.3	0.8	8.3	0.8	8.3
10	8.3	0.7	9.0	0.7	9.0	0.7	9.0	0.7	9.0
12	10.0	0.5	10.5	0.5	10.5	0.5	10.5	0.5	10.5
14	11.7	0.4	12.1	0.4	12.1	0.4	12.1	0.4	12.1
16	13.3	0.3	13.6	0.3	13.6	0.3	13.6	0.3	13.6
20	16.7	0.2	16.9	0.2	16.9	0.2	16.9	0.2	16.9
24	20.0	0.1	20.1	0.1	20.1	0.1	20.1	0.1	20.1
28	23.3	0.1	23.4	0.1	23.4	0.1	23.4	0.1	23.4
32	26.7	0.1	26.8	0.1	26.8	0.1	26.8	0.1	26.8

SOURCE: Table 50.1 from AWWA C-150.

If anticipated surge pressures are greater than 100 lb/in^2, which results from instantaneous stoppage of a column of water moving at 2 ft/s, then the actual anticipated pressures must be used.

$$P_i = 2.0 \, (200 + 100)$$
$$= 600 \, \text{lb/in}^2$$

$$t = \frac{P_iD}{2S} = 600 \times 32.00/2 \times 42,000 = 0.23 \text{ in}$$

Net thickness t for internal pressure is 0.23 in.

TABLE 5.25   Designs Values for Standard Laying Conditions

Laying Condition*	Description	E'	Bedding angle, degrees	K_b	K_s
Type 1[†]	Flat-bottom trench[‡] Loose backfill.	150	30	0.235	0.108
Type 2	Flat-bottom trench. Backfill lightly consolidated to centerline of pipe.	300	45	0.210	0.105
Type 3	Pipe bedded in 4-in-minimum loose soil.[§] Backfill lightly consolidated to top of pipe.	400	60	0.189	0.103
Type 4	Pipe bedded in sand, gravel, or crushed stone to depth of ⅛ pipe diameter, 4-in minimum. Backfill compacted to top of pipe. (Approx. 80 percent standard proctor, AASHTO T-99)[¶]				
Type 5	Pipe bedded to its centerline in compacted granular material, 4-in minimum under pipe. Compacted granular or select[§] material to top of pipe. (Approx. 90 percent standard proctor, AASHTO T-99)[¶]	700	150	0.128	0.085

* See Fig. 5.15
[†] For pipe 30 in and larger, consideration should be given to the use of laying conditions other than type 1.
[‡] Flat-bottom is defined as "undisturbed earth."
[§]Loose soil or select material is defined as "native soil excavated from the trench, free of rocks, foreign material, and frozen earth."
[¶] AASHTO T-99, "Moisture Density Relations of Soils Using a 5.5 lb (2.5 kg) Rammer 12-in (305-mm) Drop."
SOURCE: Table 50.2 from AWWA C-150.

TABLE 5.26   Allowances for Casting Tolerance

Size, in	Casting tolerance, in
3–8	0.05
10–12	0.06
14–42	0.07
48	0.08
54	0.09

SOURCE: Table 50.3 from AWWA C-150.

TABLE 5.27 Reduction Factors _R_ for Truck Load Calculations

Size, in	\< 4	4–7	8–10	\> 10
	Depth of cover, _ft_			
	Reduction factor			
3–12	1.00	1.00	1.00	1.00
14	0.92	1.00	1.00	1.00
16	0.88	0.95	1.00	1.00
18	0.85	0.90	1.00	1.00
20	0.83	0.90	0.95	1.00
24–30	0.81	0.85	0.95	1.00
36–54	0.80	0.85	0.90	1.00

SOURCE: Table 50.4 from AWWA C-150.

TABLE 5.28 Standard Thickness Classes of Ductile-Iron Pipe

Size, _in_	Outside diameter, _in_	50	51	52	53	54	55	56
		Thickness class						
		Thickness, _in_						
3	3.96	—	0.25	0.28	0.31	0.34	0.37	0.40
4	4.80	—	0.26	0.29	0.32	0.35	0.38	0.41
6	6.90	0.25	0.28	0.31	0.34	0.37	0.40	0.43
8	9.05	0.27	0.30	0.33	0.36	0.39	0.42	0.45
10	11.10	0.29	0.32	0.35	0.38	0.41	0.44	0.47
12	13.20	0.31	0.34	0.37	0.40	0.43	0.46	0.49
14	15.30	0.33	0.36	0.39	0.42	0.45	0.48	0.51
16	17.40	0.34	0.37	0.40	0.43	0.46	0.49	0.52
18	19.50	0.35	0.38	0.41	0.44	0.47	0.50	0.53
20	21.60	0.36	0.39	0.42	0.45	0.48	0.51	0.54
24	25.80	0.38	0.41	0.44	0.47	0.50	0.53	0.56
30	32.00	0.39	0.43	0.47	0.51	0.55	0.59	0.63
36	38.30	0.43	0.48	0.53	0.58	0.63	0.68	0.73
42	44.50	0.47	0.53	0.59	0.65	0.71	0.77	0.83
48	50.80	0.51	0.58	0.65	0.72	0.79	0.86	0.93
54	57.10	0.57	0.65	0.73	0.81	0.89	0.97	1.05

SOURCE: Table 50.5 from AWWA C150

TABLE 5.29   Diameter-Thickness Ratios for Laying Condition Type 2*

Trench load (P_v), lb/in^2			Trench load (P_v), lb/in^2		
Bending stress design	Deflection design	D/t or D/t_1	Bending stress design	Deflection design	D/t or D/t_1
6.29	6.18	170	8.99	7.46	128
6.34	6.19	169	9.07	7.51	127
6.39	6.21	168	9.16	7.57	126
6.44	6.23	167			
6.50	6.25	166	9.25	7.63	125
			9.33	7.69	124
6.55	6.26	165	9.42	7.75	123
6.60	6.28	164	9.51	7.81	122
6.66	6.30	163	9.60	7.87	121
6.71	6.32	162			
6.77	6.34	161	9.70	7.94	120
			9.79	8.01	119
6.82	6.37	160	9.89	8.08	118
6.88	6.39	159	9.99	8.16	117
6.94	6.41	158	10.09	8.23	116
6.99	6.43	157			
7.05	6.46	156	10.19	8.31	115
			10.29	8.40	114
7.11	6.48	155	10.40	8.48	113
7.17	6.50	154	10.51	8.57	112
7.23	6.53	153	10.62	8.66	111
7.29	6.56	152			
7.35	6.58	151	10.73	8.76	110
			10.84	8.86	109
7.42	6.61	150	10.96	8.96	108
7.48	6.64	149	11.08	9.07	107
7.54	6.67	148	11.21	9.18	106
7.61	6.70	147			
7.67	6.73	146	11.33	9.29	105
			11.46	9.41	104
7.74	6.76	145	11.59	9.54	103
7.80	6.79	144	11.73	9.67	102
7.87	6.83	143	11.87	9.80	101
7.94	6.86	142			
8.01	6.89	141	12.01	9.94	100
			12.16	10.09	99
8.08	6.93	140	12.31	10.24	98
8.15	6.97	139	12.46	10.40	97
8.22	7.01	138	12.62	10.56	96
8.29	7.05	137			
8.37	7.09	136	12.79	10.73	95
			12.96	10.91	94
8.44	7.13	135	13.13	11.10	93
8.52	7.17	134	13.31	11.29	92
8.59	7.22	133	13.49	11.50	91
8.67	7.26	132			
8.75	7.31	131	13.68	11.71	90
			13.88	11.94	89
			14.08	12.17	88
8.83	7.36	130	14.30	12.42	87
8.91	7.41	129	14.51	12.67	86

*See Fig. 5.15.
SOURCE: Table 50.8 from AWWA C-150

3. Selection of net thickness and addition of allowances. The larger of the thicknesses is given by the design for trench load, step 1, and 0.25 in is selected.

Net thickness = 0.25 in

Service allowance = 0.08 in

Minimum thickness = 0.33 in

Casting tolerance = 0.07 in

Total calculated thickness = 0.40 in

4. Selection of standard thickness and class. The total calculated thickness of 0.40 in is nearest to 0.39, class 50, in Table 5.28. Therefore, class 50 is selected.

Bibliography

1. American Association of Civil Engineers and Water Pollution Control Federation, Gravity Sanitary Sewer, "Design and Construction," 1982.
2. American Iron and Steel Institute, Handbook of Steel Drainage and Highway Construction Products, Donnelley, New York, 1971.
3. AWWA Standards: M-11, M-9, M-23, C-150, C-200, C-206, C-300, C-301, C-303, C-400, C-401, C-402, C-403, C-900, C-901, C-905, and C-950, American Water Works Association, Denver, Colo.
4. Bishop, R. R., "Courses Notebook," Utah State University, Logan, Utah, 1983.
5. Concrete Pipe Division of U.S. Pipe and Foundary Company, "Bulletin 200," Birmingham, Ala. (no date).
6. Devine, Miles, "Courses Notebook," Utah State University, Logan, Utah, 1980.
7. Ductile Iron Pipe Research Association, Thrust Restraint Design for Ductile Iron Pipe," Birmingham, Ala., 1984.
8. Howard, Amster K., "Modulus of Soil Reaction (E') Values for Buried Flexible Pipe," J. Geotech. Eng. Div., ASCE, vol. 103, no. GT, proceedings paper 127000, January 1977.
9. Moser, A. P., John Clark, and D. P. Blair, "Strains Induced by Combined Loading in Buried Pressurized Fiberglass Pipe," Proc. ASCE International Conference on Advances in Underground Pipeline Engineering, ASCE, Madison, Wis., 1985.
10. Moser, A. P., "Course Notebook," Utah State University, Logan, Utah, 1983.
11. Moser, A. P., R. K. Watkins, and O. K. Shupe, "Design and Performance of PVC Pipes Subjected to External Soil Pressure," Buried Structures Laboratory, Utah State University, Logan, Utah, 1976.
12. Piping Systems Institute, "Course Notebook," Utah State University, Logan, Utah, 1980.
13. Spangler, M. G., and R. L. Handy, "Soil Engineering," Intext Educational Publ., New York, 1973.
14. Timoshenko, S. P., and J. M. Gere, Theory of Elastic Stability, 2d ed., McGraw-Hill, New York, 1961.
15. Uni-Bell PVC Pipe Association, Handbook of PVC Pipe Design and Construction, Dallas, Tex., 1982.
16. Walker, Robert P., "Courses Notebook," Utah State University, Logan, Utah, 1983.

Index

About the Author

A. P. Moser is head of the Mechanical Engineering
Department at Utah State University. He is the founder
and director of the Piping Systems Institute, which offers a
course conducted by leading authorities for consulting
engineers who deal with pipe and piping systems. He has
published and presented over 30 papers on buried pipe
design and installation and has written numerous
technical reports on the subject. He has served as a
consultant to such companies as Owens-Corning Fiberglass
Corporation, U.S. Pipe and Foundry, Intermountain
Technology, Inc., and Johns-Mansville Corporation. Dr.
Moser holds a Ph.D degree in Civil Engineering from the
University of Colorado and M.S. and B.S. degrees in
Mechanical Engineering from Utah State University.